Anaerobic bacteria

Anaerobic bacteria
A functional biology

Paul N. Levett

Open University Press
Milton Keynes • Philadelphia

to Margaret

Open University Press
Celtic Court
22 Ballmoor
Buckingham MK18 1XW

and
1900 Frost Road, Suite 101
Bristol, PA 19007, USA

First Published 1990

British Library Cataloguing in Publication Data

Levett, P. N. (Paul Nigel) *1957–*
    Anaerobic bacteria.
    1. Anaerobic bacteria
    I. Title
    589.9

    ISBN 0 335 09205 5 pbk
         0 335 09206 3

Library of Congress Cataloging-in-Publication Number Available

Typeset by Scarborough Typesetting Services
Printed in Great Britain by St Edmundsbury Press Ltd
Bury St Edmunds, Suffolk

Contents

Preface

Anaerobic bacteria are ubiquitous, yet until comparatively recently the popular conception of anaerobes was largely restricted to those few organisms causing the classical human diseases of botulism, tetanus and gas gangrene. Major technical advances in cultural methods have facilitated a vast increase in the study of anaerobes from all sources. Initially advances were made in the fields of rumen ecology and biochemistry, but it soon became apparent that non-spore-forming anaerobes were an important cause of infection in man and animals. More recently, these techniques have been applied to the study of obligate anaerobes from environmental sources, such as the methanogens and the sulphate-reducing bacteria. The recent detection of anaerobes in hitherto unsuspected habitats such as hypersaline lakes is further evidence of their diverse environmental distribution. The application of chemotaxonomic methods to identification and classification has similarly revolutionized the taxonomy of anaerobic bacteria.

The aims of this short text are to provide a brief introduction to the diversity and importance of anaerobic bacteria and to outline some aspects of anaerobic metabolism. In addition, a short description of anaerobic methodologies is included.

There are many people to whom I am grateful for their help and encouragement. I wish to thank Lynda Petley for first stimulating my interest in anaerobes, Dr A. Trevor Willis both for his instruction and his enthusiasm for the subject, my colleagues at the Queen Elizabeth Hospital, including Emerson Haines who produced the illustrations, and lastly the librarians who have facilitated my reading, in particular Elaine Urquhart at the University of Ulster. Finally, I wish to thank Richard Baggaley of Open University Press for his support and, not least, patience.

1

Introduction: what is an anaerobe?

The existence of anaerobic organisms has been recognized since the observations of Antonie van Leeuwenhoek in the seventeenth century, but it was not until the work of Louis Pasteur on the butyric fermentation and his demonstration that the organism responsible grew only in the absence of air that the concept of anaerobiosis was established. Indeed, Pasteur wrote:

> there appears to exist a class of beings capable of living without air by obtaining oxygen from certain organic substances which undergo a slow and progressive decomposition during the process of their utilization.

This work was followed shortly after by Pasteur's description of *Clostridium septicum* (*Vibrion septique*) as a cause of gas gangrene. Thus anaerobic bacteriology is as old as the history of microbiology itself.

Until recent years only those organisms of obvious medical, veterinary or industrial importance were extensively studied. While the study of organisms for purely academic reasons may not be fashionable, the changing needs of man, particularly in the rapidly expanding area of biotechnology, render this attitude inappropriate. Only by the study of unusual, and sometimes difficult to work with, will organisms with novel characteristics and mechanisms that may be of commercial or ecological value be discovered.

A recurring theme throughout this volume is the observation that in many ways obligate anaerobes are very similar to aerobic and facultatively anaerobic bacteria. Anaerobic organisms differ from aerobes in their relationships with oxygen and this has an obvious effect upon their metabolism. However, even in the area of energy metabolism many pathways are common to both aerobes and anaerobes (see Chapter 5).

What is an anaerobe?

When it was first formed, about 4.5×10^9 years ago, the Earth's atmosphere was devoid of oxygen. It was almost certainly composed of a highly reduced combination of methane, ammonia, hydrogen and water. The first pro-karyotic life forms appeared approximately 3.5 x 10^9 years ago, according to the evidence of microfossils found in sedimentary deposits in Australia, and the presence of stromatolites. Stromatolites are fossilized bacterial mats or colonies of bacteria embedded with minerals. Modern-day stromatolites contain cyanobacteria, which were the first oxygenic photosynthetic organisms to evolve about 2 x 10^9 years ago. The bacteria within the oldest stromatolites were probably photosynthetic Eubacteria (see below) similar to the genus *Chloroflexus*.

It is probable that the first prokaryotes were Archaebacteria (see Chapter 3). These organisms were most likely obligate anaerobes which derived their energy from sulphur reduction. Similarly, phylogenetic evidence suggests that the ancestral Eubacteria were thermophilic, and possibly photosynthetic, anaerobes. This hypothesis is in accord with the evidence derived from stromatolites.

Initially the highly reduced surface of the Earth delayed the appearance and subsequent increase of oxygen in the atmosphere. It was not until the terrestrial reservoirs of reducing substances were exhausted that the oxygen content of the atmosphere began to increase to the significant level encountered today. This occurred comparatively recently, about 640 million years ago. Oxygen-ation of the oceans at this time resulted in the deposition of oxidized iron, with the production of the banded-iron formations we see today. Thus for at least 1.5 x 10^9 (1500 million) years life on Earth was exclusively anaerobic. For a further 1.4 x 10^9 years there was much less oxygen available than is presently so.

It is hardly surprising therefore that there exists an enormous diversity among those organisms considered to be anaerobes. The obligately anaerobic bacteria are not a homogeneous group of organisms. The extent of the diversity exhibited by anaerobic bacteria is exemplified in Chapter 3.

A recurrent problem for microbiologists who encounter anaerobes is that of defining exactly what anaerobes are. While this may seem an easy task, a considerable number of attempts have been made to provide academically sound yet practically workable definitions. The problems associated with the definition of an anaerobe were cogently discussed by Morris (1975). He also noted that all obligate anaerobes are characterized by two attributes:

(i) they generate energy and synthesize biomass without use of molecular oxygen;
(ii) their sensitivity to oxygen is such that they cannot grow in air (containing 20% v/v oxygen).

Confusion has existed over the use of adjectives such as 'aerotolerant', 'fastidious' and 'strict', as applied to anaerobes, since they have invariably meant one thing to one author but quite another to the next. Without labouring the point, it is abundantly clear that any volume on anaerobes must contain a clear, unambiguous definition of what exactly is meant by this vocabulary. Moreover, such a set of definitions must be closely adhered to if confusion is not to be the result.

Throughout this volume, the now customary distinction will be made between obligate anaerobes, facultative anaerobes and obligate aerobes:

(i) *Obligate anaerobes* are those organisms which cannot grow in the presence of atmospheric oxygen. A suitable practical definition, derived from the study of clinically important anaerobes, is that an obligate anaerobe will not grow on the surface of an agar plate incubated aerobically, assuming other conditions such as medium and incubation temperature are appropriate. Moreover, obligate anaerobes are almost invariably inhibited by metronidazole (see Chapter 2).

(ii) *Facultative anaerobes* are organisms which grow under both aerobic and anaerobic conditions. Included within this category are many common organisms such as *Escherichia coli*, staphylococci and streptococci.

(iii) *Obligate aerobes* are those organisms which grow only in atmospheric air, including *Pseudomonas* and *Neisseria*.

However, there are two categories of microorganisms which may still be a cause of confusion. The first of these are the microaerophilic organisms such as *Campylobacter* spp., which require a concentration of oxygen less than that normally present in atmospheric air. These organisms are not anaerobes, despite their failure to grow in air.

A second group of bacteria which may be the cause of much practical confusion in the laboratory is those organisms which require a raised CO_2 concentration for growth (carboxyphilic or capnophilic organisms). The growth of many bacteria, including both aerobes and anaerobes, is stimulated by a raised CO_2 concentration, but capnophilic organisms have an absolute requirement for CO_2 (usually about 5–10% CO_2). On first isolation these organisms are often mistaken for anaerobes because the anaerobic atmosphere contains 10% CO_2, whereas the aerobic atmosphere does not (unless a candle jar or CO_2 incubator is used). A particular source of confusion are the CO_2-dependent streptococci, many isolates of which are mistaken for anaerobic cocci in clinical laboratories. Neither microaerophilic nor capnophilic organisms are within the scope of this text, and they are not considered further.

Within this volume, 'aerotolerant' is used to describe an obligate anaerobe which is relatively resistant to exposure to oxygen, such as *Bacteroides fragilis* or *Clostridium perfringens*. The opposite to aerotolerant would thus be an extremely oxygen-sensitive anaerobe. Strict anaerobes are those which require profoundly anaerobic conditions for growth, such as the methanogens. This in itself does not imply sensitivity to oxygen, although many strict anaerobes are also oxygen-sensitive. 'Fastidious' is used only in its nutritional context, that is in the sense of an organism having exacting nutritional requirements for growth. A further convention used throughout this volume, in order to avoid confusion, is the use of the initial capital letter 'E' when referring to the Eubacteria or 'true bacteria', as distinct from the eubacteria (the genus *Eubacterium*).

It will be apparent from the foregoing paragraphs that within the broad group of obligate anaerobes there is a considerable variation in terms of susceptibility to oxygen. The extent of variation between different anaerobes is a further theme of this text; obligate anaerobes are not a homogeneous grouping any more than are obligate aerobes, for example.

Having answered the question 'what is an anaerobe?' a further question requires consideration. 'Why are anaerobes anaerobic?' may be answered in at least two ways. To simply answer that anaerobes are anaerobic because they evolved in an anaerobic environment may seem facile, yet it is an important fact to bear in mind. When obligate anaerobes evolved there was no gaseous oxygen in the environment. Thus there was no reason for early anaerobes to process oxygen, either to utilize its oxidizing power for efficient production of energy, or to protect their cellular mechanisms from the toxic action of oxygen and the products of its reduction. With the appearance of oxygen in significant amounts in the atmosphere it became necessary for organisms to possess defensive mechanisms in order to survive in increasingly widespread and hostile aerobic environments. Organisms which evolved this capability became aerobes. This process appears to have occurred many times during the course of bacterial evolution. Many organisms have not done so, yet obligate anaerobes still occupy a vast array of habitats (see Chapter 4). The biochemical mechanisms which underlie the restriction of obligate anaerobes to life without oxygen are outlined in Chapter 5. The importance of anaerobes (and a further reason for their study) is increased by their frequent interactions with man, in both beneficial (Chapter 7) and detrimental (see both Chapters 6 and 7) ways.

References and further reading

Morris, J. G. (1975) The physiology of obligate anaerobiosis. *Advances in Microbial Physiology* **12**, 169–246.

Woese, C. R. (1987) Bacterial evolution. *Microbiological Reviews* **51**, 221–271.

2

Laboratory methods for anaerobes

Anaerobic culture methods

The concept of anaerobiosis was first defined by Louis Pasteur 130 years ago. Since that time numerous methods have been devised to remove oxygen from cultures in order to facilitate the growth of obligate anaerobes. The development of such methods was hampered initially by a lack of knowledge of the nutritional requirements of many common, but exacting, anaerobes. In addition it was not appreciated that some anaerobes are more robust in terms of oxygen tolerance than others, while there exists also considerable variation in the degree of anaerobiosis necessary in order for individual species (and strains) to initiate growth. Many of the early techniques developed for anaerobic culture are now of historical interest only. Comprehensive reviews of early anaerobic techniques were made by Hall (1929) and by Willis (1960).

Removal of oxygen from culture media may be achieved by biological, physical or chemical methods. Biological methods such as the use of actively respiring plant tissues were ineffective, but the use of other microorganisms in co-culture with obligate anaerobes is more successful. One example of this technique involves the inoculation of a non-fermentative Gram-negative rod (such as *Acinetobacter*) together with the obligate anaerobe, into media used for biochemical identification tests, on the principle that the facultative organism will not interfere with the fermentation reactions of the anaerobe. Although this technique appears in practice to be useful, it would seem wise not to use it when many better methods are available.

Another application of co-culture is in the study of interactions between obligate anaerobes and *Escherichia coli* in abdominal abscesses. Using these techniques it has been possible to investigate the effect of antibiotics upon mixed aerobic/anaerobic infections and also the action of antibiotics upon *Bacteroides fragilis*. Antibiotic action has also been studied by growing *B.*

fragilis in co-culture with mammalian tissue culture monolayers, to simulate conditions *in vivo*.

Physical methods include culture under vacuum (which fails to remove all oxygen) and replacement of the atmosphere with inert gases. The latter method as originally devised also failed to remove all oxygen, and it was not until the work of Hungate (discussed below) that the potential of this technique was realized.

A wide variety of chemical methods has been used. Many of these methods successfully attain anaerobiosis but have drawbacks which render their use undesirable. Such methods include the mixture of pyrogallic acid and sodium hydroxide, the combination of chromium and sulphuric acid, and the burning of yellow phosphorus within the vessel containing the cultures. The hazards associated with these methods are respectively, the production of carbon monoxide, the evolution of excess hydrogen (which has to be bled off from the container) and the production of phosphorus pentoxide in the container. The use of iron wool dipped in acidified copper sulphate is moderately effective and remained a popular method in some laboratories until recently.

The addition of reducing agents to cultures was also widely practised, in the form of iron filings, nails or paper clips heated in a Bunsen flame before addition to the medium prior to inoculation. Chemical reducing agents are still widely used as constituents of anaerobic culture media and are discussed below. Modern developments of chemical methods include the use of an oxygen-reducing membrane fraction derived from *E. coli*.

Anaerobic jars

The most widely used anaerobic culture techniques are those based upon the catalytic combination of oxygen and hydrogen by palladium. The introduction in 1916 of the McIntosh & Fildes anaerobic jar utilizing a palladium catalyst represented a major milestone in the development of anaerobic bacteriology, allowing the reproducible growth of obligate anaerobes as surface cultures on agar media. The McIntosh & Fildes jar was constructed of glass with a brass lid. The palladium catalyst was coated onto asbestos wool and was contained within a copper mesh envelope. This was heated in a Bunsen flame before attaching to the underside of the lid, immediately prior to closure of the jar. A vacuum was then drawn which was replaced with hydrogen. The chief hazard associated with this method is the risk of explosion when the hydrogen:oxygen ratio reaches 2:1 in the presence of the hot catalyst.

The risk of explosion was reduced considerably by the introduction of electrically heated catalyst. After sealing of the jar and addition of hydrogen,

Fig. 2.1 Anaerobic jars. From left to right, glass McIntosh & Fildes jar; modified McIntosh & Fildes jar using electrically heated catalyst, constructed of brass; standard BTL jar; modern polycarbonate jar with sachets of 'cold' catalyst.

the catalyst was heated by attaching terminals on the lid to an electrical supply. This method, and modifications of it, remained in wide use for almost 40 years. In 1954 the safety of anaerobic jars was further increased by the introduction of 'cold' catalyst, operating at room temperature. The operation of anaerobic jars using room temperature catalyst has remained largely unchanged over 30 years and is described below.

The standard anaerobic jar holds 12 petri dishes and contains a sachet holding 1 g Deoxo cold catalyst (Fig. 2.1). After the jar is sealed it is connected to a vacuum pump and a mercury manometer. A vacuum is drawn to a pressure of 300 mmHg. The jar is then filled with a 90% hydrogen, 10% carbon dioxide mixture, at a low pressure (from a football bladder attached to a cylinder). The progress of catalysis can be confirmed as the lid of the jar becomes warm as the catalyst acts. After 15 minutes the manometer is reconnected to the jar; if the catalyst has functioned correctly and if the jar has no leaks, a secondary vacuum will be present. The jar is then re-filled with the gas mixture as before and it may then be placed in the incubator.

There is a slight risk of explosion if the jar is opened soon after the start of catalysis, as the air admitted mixes with the remaining quantity of hydrogen

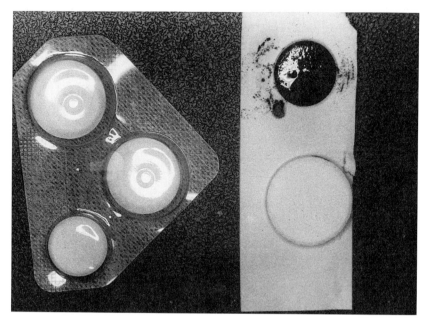

Fig. 2.2 Gas-generating tablets: sodium borohydride, tartaric acid and sodium bicarbonate (left); citric acid plus sodium bicarbonate and sodium borohydride plus cobalt chloride (right).

in the jar. For this reason some workers prefer to use a non-explosive gas mixture containing 80% nitrogen, 10% hydrogen and 10% carbon dioxide. In order to utilize this gas mixture it is necessary to remove more air from the jar initially (a vacuum of 650 mmHg is drawn). One drawback of this method is that the secondary vacuum is correspondingly smaller, so that it is difficult to monitor the operation of the catalyst by this technique.

A further modification of anaerobic jars occurred with the introduction of gas-generating envelopes. These envelopes usually contain two tablets, one of citric acid plus sodium bicarbonate and the other of sodium borohydride plus cobalt chloride (Fig. 2.2). A modified envelope contains three tablets of sodium borohydride, tartaric acid and sodium bicarbonate. Upon the addition of 10 ml water, hydrogen and carbon dioxide are evolved. Gas-generating envelopes offer several advantages over the evacuation–replacement technique. They do not require an external supply of hydrogen and carbon dioxide, nor the presence of a vacuum pump and manometer. They are thus particularly suitable for use in laboratories without a heavy demand for anaerobic cultures. Moreover, almost any robust, gas-tight container may be utilized for anaerobic culture provided a catalyst sachet can be enclosed. The drawbacks of this method are also related to the failure to remove air from the jar by vacuum pump. Since the jar contains air at atmospheric pressure, the removal of oxygen takes a longer time than when using the evacuation–replacement technique. In addition, more hydrogen is required to ensure the removal of all the oxygen, so the volume of gas generated is always larger than necessary to fill the jar. Thus there is an over-pressure created which causes gas to be forced out of the jar through the seal between the jar and the lid. It is impossible to prevent this venting of gas or to control the proportions of the gas mixture released thereby. Therefore it has been suggested that the final proportion of carbon dioxide within the jar may sometimes be much less than the intended 10%.

Anaerobic cabinets

Anaerobic jars generate an atmosphere adequate for the growth of many strict anaerobes. However, once the jar is opened the cultures within are exposed to atmospheric oxygen tension. Many oxygen-sensitive organisms from habitats such as the large intestine, the rumen or aquatic sediments are killed by such exposure to air.

For this reason in the 1960s several anaerobic cabinets or chambers were devised. A typical modern, commercially produced cabinet is illustrated in Fig. 2.3. All anaerobic cabinets are of the same basic design. A gas-tight glove box is connected to a vacuum pump and a cylinder containing 80% nitrogen, 10% hydrogen and 10% carbon dioxide. An appropriate amount of cold catalyst is included within the chamber, which is flushed through several times with the anaerobic gas mixture. Anaerobiosis is normally attained after 12–18 hours, depending upon the size of the chamber and the extent to which air remains in the chamber after flushing with the gas mixture. The chamber is maintained at a slight positive pressure to prevent the ingress of air.

Another essential feature of all anaerobic cabinets is the provision of an air-lock or interchange to allow the passage of cultures and apparatus into and out of the chamber. In most designs this takes the form of a smaller chamber from which air is withdrawn by a vacuum pump, being replaced by the anaerobic gas mixture. This evacuation–replacement cycle is normally repeated three times before the inner door is opened and material may be transferred into the inner chamber. Any residual air mixes with the inner-chamber atmosphere and oxygen is removed by catalysis.

Anaerobic cabinets vary considerably in the materials used for their construction and also in their complexity. Simple chambers may be constructed from flexible isolators to which converted anaerobic jars or even milk churns can be fitted as interchanges. Such cabinets require little sophisticated automation and can be constructed in any convenient size.

Fig. 2.3 An example of a modern anaerobic cabinet, constructed in stainless steel with a perspex front panel. Access to the chamber is via the interchange at the left-hand side of the cabinet.

Commercially available cabinets are generally of rigid construction in stainless steel and/or perspex and therefore tend to be less prone to the development of leaks than are flexible chambers. Controls are provided for the automatic operation of interchanges and topping up of gas within the chamber. Most cabinets of this type are fitted with locks which prevent the opening of the inner interchange door while the interchange is aerobic. Many commercial cabinets either contain incubators or may be operated at temperatures appropriate for incubation of cultures. A more recent refinement in some cabinets is the facility for bare-handed work; this offers great advantages over earlier designs fitted with heavy rubber gloves.

Regardless of design, all anaerobic cabinets should be used with a tray of silica gel (to absorb water following removal of oxygen), an atmospheric scrubber such as silver nitrate or activated charcoal (to remove volatile metabolites which poison the catalyst) and an Eh indicator (see the section below on indicators of anaerobiosis). The principal advantage that anaerobic cabinets offer over jars is the maintenance of anaerobiosis throughout the manipulation and incubation of cultures. It is also possible to pre-reduce media more easily using a cabinet than if anaerobic jars are used. There is no evidence that the degree of anaerobiosis (measured in terms of Eh and residual oxygen concentration) attained in cabinets is significantly greater than is achieved in properly maintained and operated jars. In well maintained cabinets the residual oxygen concentration is usually less than 20 ppm. However the provision of continuous anaerobiosis is essential if some organisms, such as methanogens, are to be studied. Another advantage of anaerobic cabinets over other methods is the ease with which experiments may be conducted in an anaerobic atmosphere, particularly those which require the use of specific apparatus (for example enzyme purification).

The handling of large numbers of anaerobic cultures is also more conveniently done using an anaerobic chamber than a large number of jars. As the volume of anaerobic work done increases it also becomes financially beneficial to operate an anaerobic cabinet. For these reasons cabinets are becoming increasingly commonplace in clinical microbiology laboratories, and there has been much inconclusive debate as to the relative effectiveness of jars versus cabinets for the isolation of clinically important anaerobes.

Despite their convenience, there are several disadvantages associated with the use of anaerobic cabinets. It is necessary to use either electric loop-sterilizing devices or pre-sterilized disposable loops within the chamber. Neither of these options is entirely satisfactory and both are considerably more expensive than conventional nichrome wire loops. It is also extremely difficult to use a microscope within a cabinet; microscopic examination of cultures thus becomes inconvenient.

Roll-tube techniques

The study of the anaerobic bacteria and protozoa of the rumen required the development of more effective culture methods than were available 40 years ago. The need to exclude all traces of oxygen from media and to prevent exposure of cultures to air led Hungate (1950) to develop the roll-tube method of anaerobic culture. The basic techniques have since been modified by a number of workers and applied to the study of obligate anaerobes from a variety of habitats. The basis of this approach is the culture of anaerobes in broth media, or on agar media solidified in a thin layer, inside test tubes containing an anaerobic atmosphere. Thus each tube represents a sealed anaerobic environment and can therefore be handled in atmospheric air without the need for anaerobic jars or cabinets.

In order to exclude oxygen from the environment, care is taken to drive off oxygen during the preparation of media and to prevent its ingress thereafter; such media are known as pre-reduced, anaerobically sterilized (PRAS) media. One of the chief requirements of this approach is the supply of oxygen-free gas, such as carbon dioxide or nitrogen. The gas is usually passed through a vessel containing hot copper filings or chromous acid in order to remove all traces of oxygen. After the dehydrated components have been weighed and the appropriate volume of water added, the Eh indicator resazurin is added. The medium is then heated to drive off oxygen and partially reduce the ingredients. When the Eh indicator changes from pink to colourless, the reducing agent cysteine hydrochloride is added, the pH is adjusted and oxygen-free gas is allowed to bubble through the medium. Media are dispensed into tubes which are also flushed with oxygen-free nitrogen via a series of cannulas. Tubes are then sealed with rubber bungs and placed in racks with lids, prior to autoclaving. The Eh in the tubes is normally $-150\,mV$ or lower. Any tubes which develop a pink colouration after autoclaving have been oxidized and are discarded.

For inoculation of samples or transfer of cultures the anaerobic atmosphere within the tube is maintained by continuous flushing with oxygen-free nitrogen. Using roll-tube methods the most oxygen-sensitive organisms may be recovered quantitatively from the rumen, the large intestine and a variety of anaerobic environmental habitats. Roll-tube methods are considerably cheaper to use than anaerobic cabinets yet provide equally effective anaerobic conditions. The chief inhibitor to their widespread use is the degree of manual dexterity required in their manipulation.

Indicators of anaerobiosis

The successful operation of anaerobic jars may be confirmed by the measurement of a secondary vacuum before the jar is placed in the incubator. Other methods of monitoring anaerobiosis, whether biological or chemical, are retrospective in nature. Biological indicators, such as cultures of strict anaerobes, are of very little value. The growth of a strict anaerobe such as *Clostridium tetani* in a jar indicates that a good degree of anaerobiosis was obtained. However, failure of the organism to grow may be due to a number of factors other than the failure of the catalyst or a leaking jar. Similarly, growth of an obligate aerobe, such as *Pseudomonas aeruginosa*, may be due to its ability to respire anaerobically using nitrate in the culture medium as a terminal electron acceptor. Moreover, failure of an obligate aerobe to grow does not invariably indicate adequate anaerobiosis.

The chemical indicators of anaerobiosis are all Eh indicators which undergo a colour change at a certain Eh. The most widely used are those compounds which are colourless when reduced and become coloured when oxidized. The Eh at which this occurs is pH-dependent, so it is customary to quote the E_0 for a given Eh indicator (Table 2.1). The E_0 is the Eh at which 50% of the indicator molecules are oxidized and 50% are reduced; the value of E_0 at pH7 is usually quoted. The most frequently used Eh indicators are

Table 2.1 E_o' values of some redox indicators

Indicator	E_o' (mV)
Methylene blue	11
Toluidene blue	−11
Resorufin (resazurin)	−51
Cresyl violet	−167
Phenosafranine	−252
Janus green	−289
Neutral red	−359
Benzyl viologen	−385
Methyl viologen	−440

methylene blue and resazurin. Resazurin is the indicator usually used in anaerobic cabinets and in PRAS roll-tube media. A resazurin solution prepared in this way will turn pink when the oxygen concentration is about 300 ppm. For more demanding anaerobic work, either phenosafranine or benzyl viologen may be used.

Media

Culture media for anaerobes do not differ to any significant extent from media for any comparable, equally diverse group of aerobes. The chief requirement for any medium for obligate anaerobes is the exclusion of oxygen as far as possible, for the reasons discussed above. This may be achieved by the use of PRAS media in roll-tubes or by the reduction in an anaerobic cabinet or jar of conventionally prepared media in petri dishes. In either case it may be beneficial to add a reducing agent to the medium.

Reducing agents

A considerable range of reducing agents may be added to culture media, including glucose, sodium thioglycollate, cysteine hydrochloride, sodium sulphide and dithiothreitol. Some anaerobes may be inhibited by some reducing agents, so the choice of reducing agent should be made with care, particularly when attempting isolation of anaerobes from fresh samples. In addition some reducing agents, such as cysteine hydrochloride, become inhibitory on storage, so media containing them should be used as soon after preparation as possible.

Culture of specific groups of anaerobes

Methanogens are almost certainly the most exacting anaerobes to cultivate. This is related chiefly to their requirement for an extremely low Eh rather than their nutritional requirements. It is preferable to cultivate methanogens in roll-tubes because the gas mixtures used for cultivation contain either 80% hydrogen and 20% carbon dioxide, or 50% hydrogen and 50% carbon dioxide. The use of such high concentrations of hydrogen in anaerobic cabinets carries with it the risk of explosion.

Isolation of methanogens may be accomplished by enrichment in minimal nutrient media containing hydrogen, formate or methanol as the electron donor. Pure cultures are obtained by serial dilution of the enrichment culture in agar roll-tubes. Most methanogens, with the obvious exception of the thermophilic species, have optimum growth temperatures either within the range 20–25°C or between 36 and 40°C, depending upon their habitat. Methanogens may be detected by the characteristic fluorescence of co-enzyme F_{420} under ultraviolet light.

The use of roll-tube methods is also necessary when attempting quantitative recovery of organisms from the normal flora of animals and man. Most of the anaerobes present in the normal flora are not extremely oxygen-sensitive, but a small proportion are.

Two approaches can be adopted in flora studies:

(i) to isolate all components of the flora with a shared characteristic, such as cellulolytic activity;
(ii) to isolate specific taxonomic groups regardless of their metabolic activity.

The former approach is used when performing ecological studies, whether of normal flora or environmental habitats. Examples of this approach include linseed oil enrichment for *Anaerovibrio* spp. from the rumen, nitrogen-free media to recover nitrogen-fixing clostridia from soil, and the use of egg-yolk agar to isolate lecithinase and lipase-producing organisms from clinical material.

The latter approach, that of selective isolation, is often used in taxonomic studies, and by clinical bacteriologists in order to isolate specific pathogens. Some of the selective agents used for this purpose are listed in Table 2.2. Physical treatments can be used to select for some anaerobic organisms. Clostridia may be isolated after either alcohol shock or heat shock has been used to kill all vegetative cells in the sample. To isolate spirochaetes, the inoculum may be placed on a 0.4 μm membrane filter overlaying the surface of an agar plate. Spirochaetes pass through the filter and spread across the plate, whereas other organisms are retained on the filter. The two strategies of

Table 2.2 Some selective agents for obligate anaerobes

Organism	Selective agent
Obligate anaerobes from clinical material	neomycin (70 mg/l) or nalidixic acid (10 mg/l)
Bacteroides spp. and *Fusobacterium* spp.	nalidixic acid (10 mg/l) + vancomycin (2.5 mg/l)
Bacteroides ureolyticus	nalidixic acid (10 mg/l) + teicoplanin (20 mg/l)
Clostridium difficile	cycloserine (250 mg/l) + cefoxitin (8 mg/l)

enrichment and selection may also be combined to make isolation techniques still more specific. The use of selective media, while helpful, does have disadvantages. Total recovery of all viable organisms in the sample is not possible unless non-selective nutrient media are also used.

In clinical bacteriology most important pathogens are relatively resistant to exposure to oxygen. A spectrum of oxygen sensitivity is still evident, however. *Clostridium perfringens* is both tolerant of exposure to oxygen in the vegetative state and is not a strict anaerobe, since it will grow in an atmosphere of up to 5% oxygen (*C. perfringens* is thus not a good organism to use as a biological indicator of anaerobiosis). *Bacteroides fragilis* is also oxygen-tolerant, withstanding exposure to atmospheric oxygen for several hours, but it is also a fairly strict anaerobe, being unable to grow on agar plates in an atmosphere containing 2% oxygen. In contrast, *C. haemolyticum* is both oxygen-sensitive and a strict anaerobe; it does not grow on agar surfaces exposed to 0.5% oxygen and vegetative cells are killed by exposure to air for 20–60 minutes.

Preservation of anaerobes

Anaerobic isolates may be preserved by regular subculture on nutrient media but this becomes extremely time consuming if many strains are involved. Clostridia may be stored for several years as cultures in cooked meat broth. Many of the more robust anaerobes survive freezing at −70°C in brain heart infusion containing 10% glycerol as a cryoprotectant. For preservation in culture collections, only freeze drying (lyophilization) is really satisfactory. Even so, many strains of anaerobes are difficult to maintain for long periods and require regular re-drying. The storage of anaerobic strains was discussed by Impey and Phillips (1984).

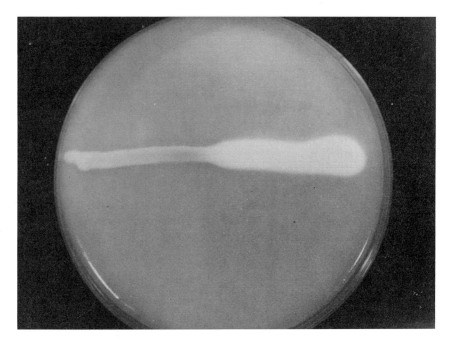

Fig. 2.4 Zone of opacity due to the production of lecithinase by *Clostridium perfringens* on egg-yolk agar. The left-hand side of the agar surface was swabbed with specific antitoxin before the plate was inoculated; this has inhibited the action of lecithinase.

Identification of anaerobes

Methods for identification of anaerobes are similar to those used for other bacteria. They differ in a few specific details which will be outlined below. Methods for anaerobe identification are covered in great detail in the laboratory manuals produced by the Virginia Polytechnic Institute (Holdeman *et al.*, 1977) and the Wadsworth Medical Center (Sutter *et al.*, 1985). The reactions of some common anaerobes in biochemical tests are shown in Chapter 3.

Conventional biochemical identification tests for anaerobes may be performed in broth media or on agar plates, and both methods are adequate for most anaerobes. Some bacteriologists prefer to use solid media for such tests so that contamination of slow-growing anaerobes may be detected by visual inspection. Some tests of course must be performed on agar plates, such as production of lecithinase on egg-yolk agar (Fig. 2.4).

Tests which detect production of acid from fermentable substrates are widely used in anaerobe identification. It is necessary to add the pH indicator, to detect acid production, after incubation of the culture since most

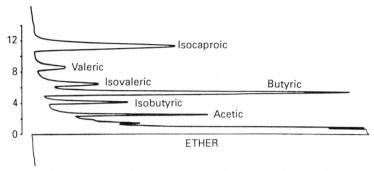

Fig. 2.5 Gas chromatogram of an ether extract of a culture of *Clostridium difficile*.

pH indicators are reduced under anaerobic conditions and change colour. Some authorities recommend the use of pH micro-electrodes to measure the reduction in pH against that in a control culture, to avoid the difficulties sometimes experienced in interpreting the colour changes of pH indicators.

Some of the problems with conventional identification techniques for anaerobes arise from the lack of saccharolytic activity shown by many species. Additional information may be gained by detecting the presence or absence of specific enzyme activities in cultures. Several identification kits based on this principle are commercially available.

Gas–liquid chromatography

Gas–liquid chromatography (GLC) has many uses in microbiology, which have been the subject of a comprehensive review (Drucker, 1981). An area in which GLC has had a great impact is the classification and identification of obligate anaerobes. The end-products of anaerobic metabolism include short-chain volatile fatty acids in the series acetic–caproic (Fig. 2.5), non-volatile acids such as lactic and succinic acids, alcohols, amines and a variety of aromatic compounds such as skatole and *p*-cresol. Some examples of the value of GLC are found in Tables 3.4 and 3.7.

References and further reading

Drucker, D. B. (1981) *Microbiological Applications of Gas Chromatography.* Cambridge: Cambridge University Press.

Hall, I. C. (1929) A review of the development and application of physical and chemical principles in the cultivation of obligately anaerobic bacteria. *Journal of Bacteriology* 17, 255–301.

Holdeman, L. V., Cato, E. P. and Moore, W. E. C. (1977) *Anaerobe Laboratory Manual*, 4th edn. Blacksburg: Virginia Polytechnic Institute and State University.

Hungate, R. (1950) The anaerobic mesophilic cellulolytic bacteria. *Bacteriological Reviews* **14**, 1–49.

Impey, C. S. and Phillips, B. A. (1984) Maintenance of anaerobic bacteria. In: *Maintenance of Microorganisms* (eds Kirsop, B. E. and Snell, J. J.), pp. 47–56. London: Academic Press.

Rosenblatt, J. E. (1986) Antimicrobial susceptibility testing of anaerobes. In: *Antibiotics in Laboratory Medicine*, 2nd edn. (ed. Lorian, V.), pp. 159–180. Baltimore: Williams & Wilkins.

Sutter, V. L., Citron, D. M., Edelstein, M. A. C. and Finegold, S. M. (1985) *Wadsworth Anaerobic Bacteriology Manual*, 4th edn. Belmont, California: Star Publications.

Willis, A. T. (1960) *Anaerobic Bacteriology in Clinical Medicine*. London: Butterworths.

3

Classification and taxonomy

Obligately anaerobic bacteria represent an extremely diverse range of organisms. Indeed, the extent of this diversity is such that anaerobes are found within both the kingdoms of Prokaryotes, the Archaebacteria and the Eubacteria. The archaebacterial anaerobes are the methanogens and a few thermophilic sulphur-metabolizers (Table 3.1), while all other obligate anaerobes are Eubacteria (Table 3.2). Many different metabolic types are represented within the anaerobic Eubacteria. Clearly it would be incorrect to regard the anaerobes as a homogeneous group of bacteria, just as it would be erroneous to consider all Gram-positive cocci in the same way. In order to make the study of anaerobic bacteria simpler and more meaningful it is usual to consider anaerobes from similar environments or those having similar metabolic characteristics together. This approach is adopted throughout this chapter.

Archaebacteria

The Archaebacteria represent an evolutionary group distinct from all other prokaryotes (the Eubacteria) and from all eukaryotic organisms. The phylogenetic evidence for the taxonomic distinction between Archaebacteria and Eubacteria was derived from 16S ribosomal RNA (rRNA) sequence characterization in the late 1970s (Woese, 1987). In this technique 16S rRNA is isolated, digested with ribonuclease T_1 and the resulting oligonucleotides are separated by two-dimensional electrophoresis. Oligonucleotides are then sequenced and sequence catalogues for different organisms are compared. Ribosomal RNA was chosen for this purpose because the ribosome is of ancient origin and is found in all living organisms. Moreover, the 16S rRNA primary structure is highly conserved, but has regions of variability. Thus both distant and relatively close relationships may be determined. Since the

Table 3.1 Anaerobic archaebacteria

Archaebacterium type	Habitat*
Methanogenic bacteria	
Order: Methanobacteriales	
Methanobacteriaceae	
Methanobacterium	A, b, d, f
Methanobrevibacter	A, b, c, d
Methanothermaceae	
Methanothermus	F
Order: Methanococcales	
Methanococcaceae	
Methanococcus	E, f, g
Order: Methanomicrobiales	
Methanomicrobiaceae	
Methanomicrobium	b, E
Methanogenium	d, E, f
Methanospirillum	a, d
Methanoplanaceae	
Methanoplanus	e
Methanosarcinaceae	
Methanosarcina	A, b, d, e
Methanococcoides	e
Methanothrix	a, d
Methanolobus	e
Thermophilic sulphur-metabolizing bacteria	
Order: Thermoproteales	
Thermoproteus	
Thermofilum	
Thermococcus	
Thermodiscus	
Desulfurococcus	
Pyrodictium	
Other genus:	
Archaeoglobus	

* a, Sewage and anaerobic digesters; b, rumen; c, non-ruminant gut; d, freshwater sediments; e, marine sediments; f, geothermal sediments; g, submarine hydrothermal vents. Capitals indicate habitats in which the genus is particularly common.

Table 3.2 Anaerobic Eubacteria

Phototrophic bacteria
Photobacteria
  Rhodospirillales
    Rhodospirillaceae
    Chromatiaceae
  Chlorobiales
    Chlorobiaceae
    Chloroflexaceae

Sulphate-reducing bacteria

Desulfobacter
Desulfobulbus
Desulfococcus
Desulfonema
Desulfosarcina
Desulfotomaculum
Desulfovibrio

Gram-positive bacteria

Bacillaceae
  Clostridium
Propionibacteriaceae
  Propionibacterium
  Eubacterium
Actinomycetaceae
  Actinomyces
  Bifidobacterium
Other Gram-positive rods
  Acetobacterium
  Lactobacillus
  Thermoanaerobacter
Anaerobic cocci
  Coprococcus
  Peptococcus
  Peptostreptococcus
  Ruminococcus
  Sarcina

Gram-negative bacteria

Bacteroidaceae
  Bacteroides
  Fusobacterium
  Leptotrichia
Haloanaerobiaceae
  Haloanaerobium
  Halobacteroides
  Sporohalobacter
Spirochaetaceae
  Spirochaeta
  Treponema
Other Gram-negative rods
  Acetivibrio
  Acetofilamentum
  Anaerobiospirillum
  Anaerovibrio
  Butyrivibrio
  Lachnospira
  Mobiluncus
  Pectinatus
  Selenomonas
  Succinimonas
  Succinivibrio
  Syntrophomonas
  Wolinella
  Zymophilus
Anaerobic cocci
  Acidaminococcus
  Megasphaera
  Veillonella

Mycoplasmas

  Anaeroplasma

molecule is relatively large, over 1500 nucleotides in size, the results obtained are statistically more acceptable than if a smaller molecule such as the 5S rRNA were used.

Comparison of 16S rRNA catalogues indicates that Archaebacteria and Eubacteria arose from a common ancestor some three billion years ago and are no more closely related to each other than either group is to the eukaryotes. In addition to this fundamental phylogenetic distinction between Archaebacteria and Eubacteria there are a number of striking phenotypic differences between these two groups of bacteria:

(i) Eubacterial cell walls contain muramic acid, whereas those of Archaebacteria do not;
(ii) membrane lipids in Eubacteria are glycerol esters of fatty acids; Archaebacterial lipids are diethers of glycerol and isoprenoids;
(iii) Archaebacteria possess tRNA devoid of ribothymidine in the TΨC loop;
(iv) Archaebacterial RNA polymerases have distinct subunit structures.

Within the Archaebacteria there are three distinct phenotypic groups, the largest and most significant of which is the methanogens (Table 3.1). The remaining two groups are the extreme halophiles of the order Halobacteriales and the extremely thermophilic sulphur-metabolizing bacteria. All methanogens are obligate anaerobes as are some of the thermophilic sulphur-metabolizers (Table 3.1).

Methanogenic bacteria

The methanogens are metabolically restricted, yet they occupy an extremely diverse range of habitats. Methanogenesis occurs in all anaerobic environments where organic compounds are degraded. These include the rumen and other gastrointestinal tract habitats, marine and freshwater sediments, geothermal springs and undersea hydrothermal vents (see Chapter 4). The industrial use of methanogens in the digestion of organic wastes is discussed in Chapter 7.

Research on anaerobic habitats is continually revealing new methanogenic species, so any description of methanogen taxonomy requires frequent revision. Methanogens may be divided into three orders on the basis of 16S rRNA homology (Table 3.1). The Methanobacteriales are all rod-shaped organisms, the cell envelopes of which contain pseudomurein. These are the most metabolically restricted methanogens, most species being able only to use H_2 as a substrate for methanogenesis; a few species can also utilize formate. The major habitats of the Methanobacteriaceae are the rumen and the gastrointestinal tracts of other animals, where there is less competition from sulphate-reducing bacteria for H_2. This may explain the inability of these methanogens to utilize other substrates.

Within the order Methanococcales there is one genus, *Methanococcus*. All methanococci have protein cell envelopes and most can use H_2 and formate as substrates for methanogenesis. All species of *Methanococcus* have been isolated from marine habitats. The order Methanomicrobiales is the most heterogeneous group of methanogens. The genus *Methanogenium* is composed of cocci which contain protein or glycoprotein in their cell walls. Organisms of this genus metabolize H_2 and formate, and their major habitats are aquatic sediments. The Methanosarcinaceae is a family comprising four genera of cocci, with protein, glycoprotein or heteropolysaccharide cell envelopes. They are also the most metabolically versatile methanogens, many species utilizing H_2, methanol, acetate and methylamines as methanogenic substrates.

Thermophilic sulphur-metabolizing bacteria

These Archaebacteria are classified within the order Thermoproteales. They are obligately thermophilic, optimum temperatures ranging from 85–105°C, and acidophilic, with pH optima from 5.5 to 6.8. Their metabolism is varied. *Thermoproteus* spp. are chemoautotrophs, gaining energy by reduction of sulphur to hydrogen sulphide. One species, *T. tenax*, is also capable of sulphur respiration of organic compounds such as glucose and ethanol. Other genera (*Thermofilum*, *Thermococcus* and *Desulfurococcus*) are also chemoheterotrophs using sulphur respiration. *Thermodiscus* and *Pyrodictium* are two genera of disc-shaped bacteria isolated from submarine hydrothermal vents. *Thermodiscus* are sulphur-respiring heterotrophs while *Pyrodictium* spp. have an autotrophic metabolism similar to *Thermoproteus*.

Phototrophic bacteria

The anaerobic phototrophic bacteria are Gram-negative organisms which inhabit anaerobic aquatic environments rich in sulphides, including stagnant water, salt lakes, estuaries and sulphur springs. They are classified within the suborder Anoxyphotobacteria and comprise two orders, the Rhodospirillales (purple phototrophic bacteria) and the Chlorobiales (green phototrophic bacteria). Within the order Rhodospirillales there are two families, the Rhodospirillaceae (the purple non-sulphur bacteria, formerly termed the Athiorhodaceae) and the Chromatiaceae (the purple sulphur bacteria, formerly the Thiorhodaceae). Similarly, there are two families within the Chlorobiales, these being the Chlorobiaceae and the Chloroflexaceae.

The phototrophic anaerobes are a heterogeneous group, the members of which share one common attribute, that of growing phototrophically without the use of water as an electron donor. Some cyanobacteria may also grow phototrophically using H_2S as an electron donor when H_2S inhibits their aerobic photosystem. These organisms are thus regarded as facultatively anaerobic phototrophs and will not be considered further.

The distinction between the Rhodospirillales and the Chlorobiales is based upon their internal morphology and the nature of their photosynthetic pigments. The purple phototrophs contain bacteriochlorophyll *a* or *b*, located on an intracytoplasmic membrane which is continuous with the cell membrane. They also contain carotenoids of groups 1–4 (carotenoids of the spirilloxanthin, rhodopinal and okenone series). In contrast the green phototrophic bacteria contain bacteriochlorophyll *c*, *d* or *e* and carotenoids of group 5 (the isorenieratene group). The absorption spectra of photo-trophic bacteria are affected by these pigmented compounds and may be of value in identification of individual species.

Within the Rhodospirillales, most members of the Rhodospirillaceae are also capable of microaerophilic growth, but their phototrophic mechanism is inhibited by oxygen. In contrast the Chromatiaceae are strict anaerobes. A further character which distinguishes the purple non-sulphur bacteria from the remaining phototrophic anaerobes is the inability of the Rhodospi-rillaceae to utilize elemental sulphur as a donor of electrons. Moreover, the Rhodospirillaceae fail to deposit elemental sulphur as an intermediary when utilizing reduced sulphur compounds (such as H_2S) as electron donors. In contrast, the Chromatiaceae usually deposit intracellular sulphur, whereas the green photobacteria (Chlorobiales) deposit sulphur outside the cell.

The heterogeneity of the anaerobic phototrophs is further emphasized by their morphological diversity, ranging through rods of various dimensions, vibrios, filaments, spirilla and bizarre spherical shapes. The morphology of the intracytoplasmic membranes also varies considerably between groups of phototrophic anaerobes. All species within the Chlorobiales possess similar vesicular structures known as chlorosomes, which are not continuous with the cell membrane. In contrast, the intracytoplasmic membranes of the Rhodospirillales are continuous with the cell membrane, but vary widely in morphology, being vesicular, tubular or lamellar in structure.

Recent phylogenetic studies based on 16S rRNA and cytochrome *c* sequence analyses have further strengthened the view that the grouping of the anaerobic phototrophic bacteria together is an artificial one. Lipid analyses, including those of carotenoids and cellular fatty acids, have been less definitive, but it is now clear that different groups of anaerobic photobacteria may be more closely related phylogenetically to other, non-phototrophic Eubacteria than to each other.

Sulphate-reducing bacteria

Sulphate-reducing bacteria (SRB) represent a further heterogeneous group of obligate anaerobes which are conveniently considered together because of their shared ability to perform dissimilatory sulphate reduction. Dissimilatory sulphate reduction is analogous to aerobic respiration in that the sulphate ion acts as an oxidizing agent in the same way as does oxygen in the aerobic process. All SRB are strict anaerobes.

The genera of SRB have been defined on the basis of morphology rather than on physiological grounds (Table 3.3). With the exception of *Desulfonema*, all SRB are Gram-negative. The largest and most frequently encountered genera are *Desulfovibrio* and *Desulfotomaculum*. Superimposed upon this morphological classification is a further division based upon the range of organic substrates which may be utilized by a given genus (or individual species).

The first group is represented by those organisms which can oxidize lactate or pyruvate in the presence of sulphate. Generally these organisms also metabolize H_2 in the presence of CO_2 and acetate. *Desulfovibrio* spp. can also utilize a variety of other substrates, including formate, glucose, malate and choline. Growth of these group 1 organisms is usually rapid.

Group 2 comprises organisms which utilize a restricted range of substrates. Acetate is the preferred substrate for complete oxidation. Hydrogen is not utilized. Growth of these organisms is slow. *Desulfobacter* spp. are the principal representatives of this physiological group, which also includes *Desulfotomaculum acetoxidans*.

The third physiological group is composed of the more metabolically active SRB. The organisms within group 3 can all oxidize fatty acids higher than acetate. A further subdivision separates those organisms also capable of oxidizing acetate (the complete fatty acid-oxidizers; group 3.2) from those which cannot utilize acetate (incomplete fatty acid-oxidizers; group 3.1). Group 3.1 comprises *Desulfobulbus* spp. and *Desulfovibrio sapovorans*. Growth of these organisms is more rapid than those in group 2, but is not as fast as those organisms that utilize lactate and hydrogen (group 1). Group 3.2 contains the most metabolically active SRB, including *Desulfococcus*, *Desulfosarcina* and *Desulfonema*. A wide variety of substrates may be utilized, including fatty acids, H_2, formate, alcohols, succinate, fumarate and aromatic carboxylic acids. Growth of these organisms is slower than that of other SRB.

The classification of SRB on morphological criteria is not supported by a study of guanine to cytosine (G:C) ratios. Within the largest genus *Desulfovibrio*, G:C ratios range from 49 to 65 mol%. A similar range exists within *Desulfotomaculum* (37–49 mol%). Clearly, such a range is incompatible with generic homogeneity.

Table 3.3 Classification of sulphate-reducing bacteria

Genus	Morphology	Motility	Flagella	Spores	Desulfoviridin	Cytochrome	Physiological group
Desulfovibrio	vibrios	+	single, polar	−	+	c_3	1
Desulfotomaculum	rods	+	peritrichous	+	−	b	1
Desulfobacter	rods/cocco-bacilli	−	−	−	−	b,c	2
Desulfococcus	cocci	−	−	−	−	b,c	3:2
Desulfosarcina	irregular, in packets	−	−	−	−	*	3:2
Desulfobulbus	citron-shaped cocci	−	−	−	−	b,c	3:1
Desulfonema	filaments	+	−	−	−	b,c	3:1

* Not determined.

Spore-forming rods

The remaining Eubacterial anaerobes are all chemoheterotrophs and include all the anaerobes of medical and veterinary importance as well as many of industrial and ecological significance. These organisms can be divided conveniently by their morphology and Gram reactions.

With the exception of the sulphate-reducing genus *Desulfotomaculum* and the obligately-halophilic *Sporohalobacter*, all anaerobic, spore-forming rods are included within the genus *Clostridium*. This is a very large genus, currently comprising over 100 validly described species. Once again, this is a very diverse collection of species, the lumping together of which within a single genus does not seem justified by phylogenetic evidence. The type species, *C. butyricum*, has a G:C ratio of 27 mol% while that for the genus ranges from 22 to 55 mol%. The majority of *Clostridium* spp. have G:C ratios less than 30 mol%, but the remaining twenty or so species have G:C ratios which occupy a continuous progression from 30 to 55 mol%. While it is apparent that organisms with G:C ratios of 27 and 55 mol% are unrelated, it is not easy to identify discrete groups in between these extremes. DNA hybridization methods, which could be helpful in this respect, have not been applied to the genus as a whole.

This picture is further complicated if numerical methods are applied. The groups formed by such analysis do not separate species with high G:C ratios from the remainder, nor do they correspond with the groups distinguished by DNA hybridization.

The classification of clostridia was for many years confused. The problems of clostridial taxonomy were discussed eloquently by Willis (1969) and more recently by Hobbs (1986). The scheme proposed by Prevot divided the anaerobic spore-bearing bacilli into nine genera, distributed between four families. The scheme was not widely accepted and the genus *Clostridium* remains a convenient, if scientifically unsatisfactory, repository for anaerobic spore-formers. It is possible to provide a working definition of the genus while recognizing that exceptions to this description are common. *Clostridium* spp. are obligately anaerobic, Gram-positive bacilli which form resistant endospores. Members of this genus ferment a wide range of carbohydrates and amino acids. The habitats of these organisms include soil and the intestinal tracts of man and animals. They are found widely distributed in foods of both plant and animal origin. Several species are pathogenic for animals and man. The pathogenic species produce protein-aceous exotoxins. The end-products of metabolism, determined by GLC, are varied. Many species produce primarily acetic and butyric acids, while others produce a range of short-chain fatty acids and alcohols. The characteristics of some common clostridia are shown in Table 3.4.

Table 3.4 Characteristics of some common clostridia

Species	Spore location*	β-haemolysis	Lecithinase	Lipase	Indole	Acid produced from:				Hydrolysis of:		End-products of glucose metabolism**
						Glucose	Lactose	Maltose	Sucrose	Gelatin	Starch	
C. acetobutylicum	ST	-	-	-	-	+	+	+	+	-	+	A,B
C. bifermentans	ST	+	+	-	+	+	-	-	-	+	-	A,p,ib,b,iv,v,IC
C. botulinum (types A, B, F)	ST	+	+	+	-	+	-	-	-	+	-	A,p,ib,B,iv,v,ic
C. botulinum (types B, E, F)	ST	+	-	+	-	+	-	+	+	+	+	A,B
C. botulinum (types C, D)	ST	+	-	+	-	+	-	-	-	+	-	A,P,B
C. butyricum	ST	-	-	-	-	+	+	+	+	-	+	A,B
C. difficile	ST	-	-	-	-	+	-	-	-	-	-	A,p,ib,B,iv,v,IC
C. felsineum	ST	+	-	-	-	+	+	-	+	+	-	A,B
C. innocuum	T	-	-	-	-	+	-	-	+	-	-	A,B
C. novyi type A	ST	+	+	+	-	+	-	-	-	+	-	A,P,B
C. novyi type B	ST	+	+	+	+	+	-	-	-	+	-	A,P,B
C. paraputrificum	T	-	-	-	-	+	+	+	+	-	-	A,B
C. pasteurianum	ST	-	-	-	-	+	-	+	+	-	-	A,B
C. perfringens	ST	+	+	-	-	+	+	+	+	+	+	A,B
C. ramosum	T	-	-	-	-	+	+	+	+	-	-	A
C. septicum	ST	+	-	-	-	+	+	+	-	+	-	A,B
C. sordellii	ST	+	+	-	+	+	-	-	-	+	-	A,p,ib,b,iv,v,ic
C. sporogenes	ST	+	+	+	-	+	-	-	-	+	-	A,p,ib,B,iv,v,IC
C. tertium	T	-	-	-	-	+	+	+	+	-	-	A,B
C. tetani	T	+	-	+	+	-	-	-	-	+	-	A,B
C. tyrobutyricum	ST	+	-	-	-	+	-	-	-	-	-	A,B

* ST = sub-terminal spores, T = terminal spores.

** A = acetic, P = propionic, ib = isobutyric, B = butyric, iv = isovaleric, v = valeric, IC = isocaproic acids. Capitals indicate major end-products and lower-case letters indicate minor end-products.

Although the majority of clostridia have Gram-positive cell walls, many species appear Gram-negative, particularly when older cultures are examined. Some species, *C. clostridioforme* for example, are actually Gram-negative. A number of *Clostridium* spp. are able to grow on the surface of agar plates incubated aerobically. These species include *C. carnis*, *C. durum*, *C. histolyticum* and *C. tertium*. The growth of these species is much better under anaerobic conditions than in an aerobic atmosphere, and they do not form spores in aerobic cultures. Thus they should probably be regarded as 'facultative aerobes'. A number of clostridia rarely sporulate in the usual culture media. *Clostridium perfringens* is the most important of these species.

Gram-negative rods

This group of anaerobes comprises the Bacteroidaceae and a number of genera of uncertain taxonomic affiliation. The genera within the family Bacteroidaceae are defined by their production of end-products from glucose or peptone metabolism. *Bacteroides* spp. produce a variety of volatile and non-volatile acids. Most species produce large amounts of succinic acid, but few produce appreciable amounts of butyric acid. The genus *Fusobacterium* is characterized by the production of large amounts of butyric acid. Many species also produce acetic and lactic acids. *Leptotrichia* is a monospecific genus; *L. buccalis* produces large amounts of lactic acid as its sole end-product. The Bacteroidaceae are all inhabitants of the body cavities of man and animals, including insects. Many species of *Bacteroides* and *Fusobacterium* are pathogens of man and animals.

Differentiation of species within the genera *Bacteroides* and *Fusobacterium* is achieved by a combination of colonial and microscopic morphology and biochemical tests. Several important fusobacteria may be readily identified by a limited number of simple tests. *Fusobacterium necrophorum* produces β-haemolytic colonies on blood agar plates and produces both lipase and indole.

Bacteroides spp. may be divided into three groups, each of several species, by the use of a few simple characterization tests. These groups are summarized in Table 3.5. However, recently obtained chemotaxonomic and phylogenetic evidence suggests that the genus *Bacteroides* should be confined to those bile-resistant species once referred to as the '*B. fragilis* group', with G:C ratios between 40 and 48 mol%. This group of organisms is characterized by the possession of enzymes of the hexose monophosphate shunt and pentose phosphate pathway, sphingolipids, methyl branched long-chain fatty acids, and menaquinones as the sole respiratory quinones (Shah and Collins, 1983). Proposals have been made to divide the genus on that basis (Shah and Collins, 1989; 1990). All other *Bacteroides* spp. have been

Table 3.5 Characteristics of some common *Bacteroides* species

Species	Bile resistance	Pigmented colonies	Indole	Hydrolysis of:			Acid produced from:							
				Gelatin	Aesculin	Urea	Ara	Glu	Lac	Raf	Sal	Sta	Suc	Tret
B. asaccharolyticus	−	+	+	+	−	−	−	−	−	−	−	−	−	−
B. bivius	−	−	−	+	−	−	−	+	+	−	−	+	−	−
B. disiens	−	−	−	+	−	−	−	+	−	−	−	+	−	−
B. distasonis*	+	−	−	−	+	−	−	+	+	+	+	+	+	+
B. eggerthii*	+	−	+	+	+	−	+	+	+	−	+	+	+	−
B. fragilis*	+	−	−	−	+	−	−	+	+	+	−	+	+	+
B. hypermegas	+	−	−	−	−	−	+	+	+	+	−	−	+	+
B. intermedius	−	+	+	+	+	−	−	+	−	−	+	+	+	+
B. levii	−	+	−	+	−	−	−	+	+	−	−	−	−	−
B. melaninogenicus	−	+	−	+	+	−	−	+	+	−	−	+	−	−
B. oralis	−	−	−	+/−	+	−	−	+	+	+	+	+	+	−
B. ovatus*	+	−	+	+	+	−	+	+	+	+	+	+	+	+
B. ruminicola	−	−	−	−	+	−	+	+	+	+	−	−	+	−
B. splanchnicus	+	−	+	+	+	−	+	+	−	−	−	+	−	−
B. thetaiotaomicron*	+	−	+	−	+	−	+	+	+	+	+	+	+	+
B. uniformis*	+	−	+	−	+	−	+	+	+	+	+	−	+	−
B. ureolyticus**	−	−	−	−	−	+	−	−	−	−	−	−	−	−
B. vulgatus	+	−	−	+	+	−	+	+	+	+	−	+	+	−

† Ara = arabinose, Glu = glucose, Lac = lactose, Raf = raffinose, Sal = salicin, Sta = starch, Suc = sucrose, Tre = trehalose.
* These species are members of the 'B. Fragilis group'.
** Produces pitting colonies on agar surfaces.

Table 3.6 Recently created genera comprised of species transferred from the genus *Bacteroides*

Anaerorhabdus furcosus
Fibrobacter succinogenes
Megamonas hypermegas
Mitsuokella multiacidus
Porphyromonas asaccharolyticus
　P. endodontalis
　P. gingivalis
Prevotella
Rikenella microfusus
Ruminobacter amylophilus
Sebaldella termiditis
Tissierella praeacutus

---

transferred to newly created genera (Table 3.6); it seems probable that the taxonomy of the *Bacteroides* group will remain in a state of change for some time to come.

The haloanaerobes are an interesting group of obligately halophilic Gram-negative rods isolated from hypersaline lake sediments, such as those in the Dead Sea and the Great Salt Lake in Utah, and from hypersaline waters associated with gas and oil-bearing sediments. The family Haloanaerobiaceae comprises three genera (Table 3.2), all of which are characterized by the requirement for sodium chloride concentrations in the range 0.4–2.8 M. The haloanaerobes described to date are all chemoheterotrophs, are motile by means of peritrichous flagella, and have G:C ratios around 30 mol%. *Sporohalobacter* spp. are distinguished from other haloanaerobes by their production of endospores. The end-products of carbohydrate fermentation by *Halobacteroides* and *Sporohalobacter* are acetic acid, ethanol, carbon dioxide and hydrogen, whereas *Haloanaerobium* spp. produce acetic and butyric acids, carbon dioxide and hydrogen.

Several genera of Gram-negative, acetogenic rods have recently been described. *Syntrophomonas* spp. have been isolated only in co-culture with methanogens or sulphate-reducing bacteria. They are characterized by the anaerobic β-oxidation of saturated fatty acids, using protons as electron acceptors, in the presence of H_2-consuming bacteria. *Acetofilamentum rigidum* is notable for the production of thin, straight rods which often occur as filaments of >100 μm in length. *Acetofilamentum rigidum* is fermentative, producing acetate, CO_2 and H_2 in sewage sludge. It also is inhibited by raised H_2 concentrations. *Acetivibrio* spp. have been recovered from the porcine large intestine and from methanogenic consortia. The end-products of glucose metabolism are acetic acid, ethanol, CO_2 and H_2.

Two genera of spirochaetes contain obligate anaerobes. *Spirochaeta* is a genus of chemoorganotrophs whose habitat is marine and freshwater and underlying sediments. The end-products of glucose metabolism are acetic acid, ethanol, CO_2 and H_2. The genus *Treponema* is comprised of parasites of the mucous membranes, the rumen and the large intestine. Many species have not been cultivated, but there is evidence that the human pathogens, including *T. pallidum*, are microaerophilic, requiring an oxygen concentration between 1 and 5%. However, many species are obligately anaerobic; they include several members of the oral flora (see Chapter 4) and *T. hyodysenteriae*, the cause of swine dysentery (see Chapter 6). *Treponema hyodysenteriae* grows between 36 and 42°C, has an optimum pH of 6.9 and requires a moderately reduced environment for growth (Eh less than -125 mV). The end-products of its metabolism are acetic and butyric acids, H_2 and CO_2. Other anaerobic treponemes produce acetic, propionic, butyric, lactic, formic and succinic acids.

The remaining genera of Gram-negative anaerobic rods are shown in Table 3.2. *Anaerobiospirillum* is a monospecific genus composed of helical rods; *A. succinoproducens* has bipolar tufts of flagella and exhibits corkscrew-like motility. The organism has been isolated from the faeces of dogs and humans with diarrhoea. *Mobiluncus* is a genus of motile, curved rods, the principal habitats of which appear to be the human vagina and large intestine. *Mobiluncus* spp. are important components of the flora in bacterial vaginosis (see Chapter 6). *Pectinatus* is a genus of long, slender rods originally isolated from spoiled beer. A similar genus of curved and/or helical Gram-negative rods was recently isolated from pitching yeast and was named *Zymophilus*. *Zymophilus* spp. are chemoorganotrophs, whose end-products are acetic and propionic acids. *Selenomonas* spp. and *Wolinella* spp. are inhabitants of the gingival crevice in man and animals. One species, *W. succinogenes*, is found in the rumen.

The other genera of Gram-negative rods are also components of the rumen flora. These organisms are curved or straight rods and all are highly saccharolytic. They produce a wide variety of short-chain acids from glucose metabolism. *Butyrivibrio* is the most numerous of these organisms in the rumen; similar organisms are found in the intestinal tracts of other mammals.

Gram-positive non-sporing rods

This grouping is somewhat artificial, since inclusion within it depends upon the absence of spores, yet the genera of non-sporing, Gram-positive anaerobic bacilli are not closely related. Within this group GLC is helpful for distinguishing between genera (Table 3.7).

Table 3.7 Differentiation of some Gram-positive non-sporing anaerobic rods by gas chromatography

Genus	Major end-products of glucose metabolism
Actinomyces	acetic, lactic and succinic acids
Bifidobacterium	acetic and lactic acids
Eubacterium	varied
Lactobacillus	lactic acid
Propionibacterium	acetic and propionic acids

Actinomyces spp. are Gram-positive, pleomorphic rods, which usually show branching or diphtheroid cells. Filamentous forms are common. All species produce large amounts of succinic, lactic and acetic acids. Not all strains of *Actinomyces* spp. are obligate anaerobes. Within each species there exists considerable variation in the degree of aerotolerance. Only *A. meyeri* is uniformly obligately anaerobic.

After several days' incubation, typical *Actinomyces* colonies are white, heaped-up and matt, resembling large breadcrumbs. They are usually hard and attempts to remove colonies from agar surfaces frequently result in the removal of the whole colony plus a portion of the agar medium. *Actinomyces odontolyticus* produces pink-to-brown colonies on blood agar. Growth in a fluid medium often results in the formation of macro-colonies with an appearance similar to those on solid media. The habitat of most *Actinomyces* spp. is the oral cavity of man and animals; some species produce chronic granulomatous infections in man and other animals.

The genus *Bifidobacterium* is composed of irregularly shaped, obligately anaerobic, Gram-positive rods that produce large amounts of acetic and lactic acids. *Bifidobacterium* spp. are also characterized by the production of fructose-6-phosphate phosphoketolase (F6PPK) and trypsin (Beerens *et al.*, 1986). Starch gel electrophoresis allows the differentiation of three groups within the genus, based upon electrophoretic mobility of F6PPK. Bifidobacteria are components of the large intestinal flora in man and animals. Several species (particularly *B. bifidum*, *B. infantis* and *B. longum*) are predominant in the development of the neonatal gut flora and the exclusion of potentially pathogenic organisms; these species require human milk as a growth factor. Other habitats of bifidobacteria include the human vagina, the oral cavity and sewage.

Eubacterium is a further heterogeneous genus of obligately anaerobic, pleomorphic Gram-positive rods. Microscopic morphology of *Eubacterium* spp. varies from rods resembling clostridia to small cocci. *Eubacterium* spp. may produce butyric acid as a major end-product, together with other acids, or they may produce no acidic end-products at all. Motile strains of *Eubacterium* that produce butyric acid and exhibit clostridial morphology are considered by some workers to represent asporogenous clostridia. The habitat of *Eubacterium* appears to be the intestine of man and other animals, including ruminants. Indeed, eubacteria represent one of the numerically dominant groups within the human colon (see Chapter 4).

The genus *Lactobacillus* is composed of fermentative, Gram-positive rods which produce lactic acid as their sole major end-product. Not all strains of lactobacilli are obligate anaerobes and the degree of aerotolerance exhibited is wide. However several species, such as *L. catenaforme* and *L. rogosae*, appear to be obligately anaerobic. Most strains of other *Lactobacillus* spp. grow better under anaerobic conditions than aerobically. Differentiation of species within the genus depends upon fermentation reactions, ability to grow at 45°C, production of ammonia from arginine and determination of optical rotation by the isomers of lactic acid produced (Holdeman *et al.*, 1977).

The habitats of lactobacilli include the female genital tract and the intestinal tract of man and a wide variety of animals, including rodents and birds. Some species are found in soil and on plant material. Lactobacilli are important in the food industry as components of the flora of dairy products, in particular of yoghurts and other fermented milk products (see Chapter 7).

Propionibacterium spp. are fermentative, non-motile rods which often have a diphtheroid morphology. Some species are obligately anaerobic, while others are facultative anaerobes. There are two ecological groupings of propionibacteria, one being important in cheese manufacture (see Chapter 7). The species *P. acnes*, *P. granulosum* and *P. avidum* are important components of the flora of pilosebaceous ducts and of acne lesions.

Acetobacterium is a genus of oval short rods, whose habitats include marine and fresh water sediments and sewage. These organisms perform a homoacetic fermentation of fructose and a few other substrates, oxidizing H_2 and reducing CO_2. In contrast, *Thermoanaerobacter* is a monospecific genus of thermophilic, chemoorganotrophic rods; *T. ethanolicus* was isolated from alkaline and slightly acidic hot springs in Yellowstone National Park. This species has an optimum pH between 5.8 and 8.5, but will grow over a much wider range (pH 4.4–9.8). Similarly, the optimum growth temperature is 68°C but the range of temperatures over which the organism will grow extends from 35 to 78°C. The end-products of metabolism of *T. ethanolicus* are ethanol and CO_2.

Anaerobic cocci

The genera of Gram-negative anaerobic cocci are classified within the family Veillonellaceae. There are three genera (Table 3.2). *Veillonella* spp. are parasites of mucous membranes of man and other animals. They are non-fermentative but acquire energy from the anaerobic respiration of lactate. *Acidaminococcus* is a monospecific genus of cocci inhabiting the intestinal tract of mammals: *A. fermentans* derives energy from the fermentation of amino acids; *Megasphaera* comprises large cocci (2.5 μm diameter); *M. elsdenii* is found in the rumen and ferments carbohydrates, producing a variety of acidic end-products.

The classification of Gram-positive anaerobic cocci is much less clear. A number of species are recognized on the basis of biochemical reactions and GLC end-product analyses. However, most species are biochemically inert and produce similar end-products from glucose metabolism. Moreover, many isolates appearing to be obligate anaerobes on first isolation are later found to be microaerophilic and are thus streptococci. Formerly, two genera within the family Peptococcaceae, *Peptococcus* and *Peptostreptococcus*, were recognized. Differentiation between these genera was based largely on the production of chains of cells by *Peptostreptococcus* spp. However, this characteristic is extremely variable between strains and species of both genera. Only one species of the genus *Peptococcus* (*P. niger*) is now recognized. The two genera are now differentiated by G:C ratios. Peptostreptococci have G:C ratios within the range 28–35 mol% whereas *Peptococcus niger* has a G:C ratio of 50 mol%. One species, *Peptostreptococcus anaerobius*, is readily identified by production of isocaproic acid and is a common component of anaerobic infections in both man and animals (see Chapter 6).

There are two Gram-positive genera distinct from *Peptococcus* and *Peptostreptococcus*. *Ruminococcus* comprises cellulolytic cocci which are found in large numbers in the rumen, while *Sarcina* is a genus of large cocci which form packets of eight or more cells. *Sarcina* spp. form endospores, the presence of which may be demonstrated by spore staining or by heat shock.

Mycoplasmas

One genus of anaerobic mycoplasmas has been described. *Anaeroplasma* spp. are cholesterol-requiring mycoplasmas isolated from the rumen. The end-products of metabolism include acetic, formic, lactic and succinic acids, ethanol and CO_2.

References and further reading

Beerens, H., Neut, C. and Romond, C. (1986) Important properties in the differentiation of Gram-positive non-sporing rods in the genera *Propionibacterium, Eubacterium, Actinomyces* and *Bifidobacterium*. In: *Anaerobic Bacteria in Habitats Other than Man* (eds Barnes, E. M. and Mead, G. C.), pp. 37–59. Oxford: Blackwell Scientific Publications.

Hardie, J. M. (1989) Application of chemotaxonomic techniques to the taxonomy of anaerobic bacteria. *Scandinavian Journal of Infectious Diseases* Supplement **62**, 7–14.

Hobbs, G. (1986) Some problems in the identification and taxonomy of clostridia. In: *Anaerobic Bacteria in Habitats Other than Man* (eds Barnes, E. M. and Mead, G. C.), pp. 23–36. Oxford: Blackwell Scientific Publications.

Holdeman, L. V., Cato, E. P. and Moore, W. E. C. (1977) *Anaerobe Laboratory Manual*, 4th edn. Blacksburg: Virginia Polytechnic Institute and State University.

Jones, W. J., Nagle, D. P. and Whitman, W. B. (1987) Methanogens and the diversity of archaebacteria. *Microbiological Reviews* **51**, 135–177.

Oren, A. (1986) The ecology and taxonomy of anaerobic halophilic eubacteria. *FEMS Microbiology Reviews* **39**, 23–29.

Postgate, J. R. (1984) *The Sulphate-reducing Bacteria*, 2nd edn. Cambridge: Cambridge University Press.

Shah, H. N. and Collins, M. D. (1983) Genus *Bacteroides*: a chemotaxonomical perspective. *Journal of Applied Bacteriology* **55**, 403–416.

Shah, H. N. and Collins, M. D. (1989) Proposal to restrict the genus *Bacteroides* (Castellani & Chalmers) to *Bacteroides fragilis* and closely related species. *International Journal of Systematic Bacteriology* **39**, 85–87.

Shah, H. N. and Collins, M. D. (1990) *Prevotella*, a new genus to include *Bacteroides melaninogenicus* and related species formerly classified in the genus *Bacteroides*. *International Journal of Systematic Bacteriology* **40**, 205–208.

Willis, A. T. (1969) *Clostridia of Wound Infection*. London: Butterworths.

Woese, C. R. (1987) Bacterial evolution. *Microbiological Reviews* **51**, 221–271.

Readers with an interest in taxonomy should also consult:

Bergey's Manual of Systematic Bacteriology, Vol. I (1984) (eds Krieg, N. R. and Holt, G.); Vol. II (1986) (eds Sneath, P. H. A., Mair, N. S., Sharpe, M. E. and Holt, J. G.); Vol. III (1989) (ed. Staley, J. G.). Baltimore: Williams & Wilkins.

Valid publication of new or emended taxa is effected by publication in the *International Journal of Systematic Bacteriology*. The taxonomy of anaerobes is changing rapidly, so that it is necessary to consult both *Bergey's Manual* and the *International Journal of Systematic Bacteriology* in order to keep up to date with nomenclatural changes.

4

Ecology of anaerobes in the natural environment

Despite the general susceptibility of obligate anaerobes to oxygen, the range of habitats colonized by anaerobes is wide (Table 4.1). In the gastrointestinal tracts of animals (including man) obligate anaerobes perform a number of important physiological functions. In the physical environment anaerobes play important roles in the cycling of carbon (methanogens) and sulphur (sulphate-reducing bacteria).

Table 4.1 Habitats of anaerobic bacteria

Oral cavity
Gastrointestinal tract
Female genital tract
Soil and plant tissues
Aquatic sediments
Decomposing organic matter

Oral cavity

The oral cavity of dentate animals presents a number of distinct habitats for microbial colonization. These include the buccal surfaces, the tongue, the gums, the teeth and the gingival crevices. These habitats initially become colonized soon after birth by a complex flora of which obligate anaerobes are an important component. The salivary flora is derived primarily from the tongue. As would be expected, the numbers of anaerobes on the exposed surfaces of the cheeks and the tongue are relatively low. *Veillonella* spp. represent about 10% of the flora on the human tongue and a similar proportion of the viable organisms found in saliva.

Table 4.2 Obligate anaerobes isolated from gingival crevices

Bacteroides asaccharolyticus	*Actinomyces* spp.
B. gingivalis	*Bifidobacterium* spp.
B. oralis	*Eubacterium* spp.
Bacteroides spp.	*Propionibacterium* spp.
Fusobacterium necrophorum	*Selenomonas sputigena*
F. nucleatum	*Treponema denticola*
Leptotrichia buccalis	*T. macrodentium*
Anaerobic Gram-positive cocci	

Colonization of other sites within the oral cavity occurs in a series of distinct stages, dependent upon the changing anatomy of the mouth during growth and ageing of the individual. Most research has been done on the human oral flora, but there is evidence that the general principles probably apply in other animals also. In the edentulous child there are few anaerobic habitats; after the teeth erupt the gingival crevices rapidly become colonized by a mixed flora composed of facultative and obligate anaerobes. The initial colonizers are streptococci, which adhere to the proteinaceous pellicle deposited on the teeth from the saliva. As the flora develops, the Eh in the gingival crevices falls and obligate anaerobes appear (Table 4.2).

Bacterial colonization of the teeth results in the formation of plaque. The microbial nature of plaque was first described by Antonie van Leeuwenhoek in 1683. Plaque is a three-dimensional structure consisting of bacterial cells embedded in an amorphous matrix. The matrix is composed of bacterial extracellular polymers and macromolecules derived from saliva and crevicular fluid.

Plaque first appears on uncleaned teeth after a few days. The composition of plaque varies between individuals and between sites within an individual mouth. *Streptococcus mutans* is particularly prevalent in fissure plaque, which leads to the development of carious lesions (Fig. 4.1). Approximal plaque accumulates between teeth and may be supragingival or subgingival (located in the gingival crevice). The conditions within plaque become steadily more anoxic as the plaque accumulates. Close to the surface of approximal plaque, the Eh may be $-150\,\text{mV}$, whereas in subgingival plaque it falls to $-300\,\text{mV}$. Conditions are thus favourable for the growth of extremely oxygen-sensitive anaerobes. The existence of plaque also affects the general oral environment. In individuals with good oral hygiene and no evidence of caries, saliva usually has an Eh of $+300\,\text{mV}$; in patients with caries and plaque accumulation, the salivary Eh is reduced to $+200\,\text{mV}$. Good oral hygiene decreases viable counts of oral bacteria, especially of anaerobes.

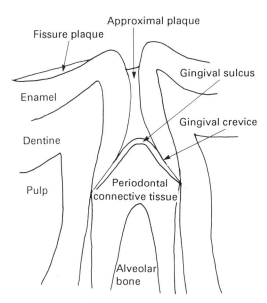

Fig. 4.1 Cross-section through teeth, showing development of plaque.

The presence of plaque is associated with the development of periodontal disease, a chronic, slowly progressive and destructive inflammatory process affecting the periodontium. Periodontal disease was a common problem in Egypt about 3500 years ago. The expression 'long in the tooth' is a reflection of the prevalence of the condition in past times. It is estimated that at least a half of all middle-aged adults have periodontal disease.

Several stages may be defined in the development of chronic periodontal disease. Plaque accumulates at the gingival margin (Fig. 4.1) and if not removed becomes calcified to form subgingival calculus. Chronic inflammation of the gingivae and periodontal membrane occurs, accompanied by degeneration of connective tissue. The gingival cuff migrates apically, revealing more of the tooth and forming periodontal pockets in which further inaccessible calculus accumulates. Eventually resorption of alveolar bone occurs leading to loosening, and finally loss of, the teeth.

In plaque associated with periodontal disease, Gram-negative organisms predominate. Anaerobes attain high numbers (approximately 10^{10} cells/g). Those detected most frequently include fusobacteria, *Bacteroides* spp., treponemes and *Selenomonas sputigena*, in addition to microaerophilic and carboxyphilic organisms such as *Campylobacter* and *Eikenella corrodoens*. Electron microscopy reveals some of the highly specific interactions between organisms in plaque. One of the most common is the presence of 'corncob' formations, filaments of *Bacterionema matruchotii* surrounded by adherent streptococci. Other specific interactions include adhesion of *Veillonella alcalescens* to *Actinomyces viscosus* and the attachment of *Bacteroides asaccharolyticus* and *Fusobacterium nucleatum* to *A. viscosus*, *A. israelii* and *A. naeslundii*.

The role of obligate anaerobes in chronic periodontal disease is not entirely clear. Black-pigmented *Bacteroides* spp., particularly *B. gingivalis*, appear to be significant because of their strong protease activity. A similar condition, known as 'broken mouth', occurs in sheep. A more acute periodontal condition known as Vincent's disease is discussed in Chapter 6.

Gastrointestinal tract

The gastrointestinal tract of many animal species represents one of the most important habitats of obligate anaerobes. In simplistic terms the gastrointestinal tract may be viewed as a tube running through the body, so that in a sense the lumen of the gut is external to the body. However, the activities of the microbial flora are of such importance to the host that the gut flora may be considered as one of the most important organs in the body.

The intestinal tract can be regarded as a semicontinuous culture system, with intermittent feeding and excretion of waste products. This system consists of a number of chambers, and the substrate (food) is moved along the system by the peristaltic action of surrounding muscles. Several laboratory-scale models have been devised using either single- or multi-chamber chemostats in order to model the ecology and biochemical activities of the gastrointestinal tract flora.

Animal gastrointestinal tracts may be divided into two broad groups:

 (i) those which are adapted for a foregut (rumen) fermentation;
 (ii) those in which a hindgut fermentation takes place.

Rumen

The foregut or rumen fermentation evolved in order to facilitate the metabolism of cellulose by herbivorous animals that do not produce cellulases. The development of this fermentation is due to the separation of the rumen from the acidic region of the stomach, the abomasum. A wide variety of animals have developed a foregut fermentation in the absence of a true rumen. These include marsupials, leaf-eating monkeys, sloths, hippopotami, camels and llamas. These animals also do not practise rumination, or 'chewing the cud'. The true ruminants include cattle, sheep, deer, gazelle, antelopes and giraffes.

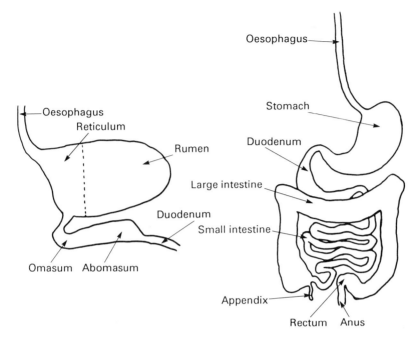

Fig. 4.2 Comparative anatomy of rumen (left) and human gastrointestinal tract (right).

The reticulo-rumen is a four-chambered modified stomach (Fig. 4.2). Liquid foods, such as milk, pass directly through the reticulum into the omasum without entering the rumen. Other foods, however, pass into the rumen where a cellulolytic microbial fermentation occurs. The volatile acids produced by this fermentation are absorbed from the rumen into the bloodstream and serve as the animal's main source of energy and carbon.

The efficiency of this digestive process is maximized by comminution of the plant material by grinding teeth and by a characteristically long retention time of the substrate within the rumen. Thus the rumen may be regarded as a semicontinuous fermentation system (Fig. 4.3).

Food particles in the rumen are mixed by muscular contractions and periodically regurgitated for further chewing, or rumination. This practice also facilitates the venting by eructation of large volumes of methane and CO_2 produced in the rumen. An average-sized cow weighing 500 kg has a rumen with a volume of about 70 litres and belches approximately 200 litres of methane and CO_2 per day. The total annual production of methane by ruminants may be greater than 2×10^8 tonnes, comparable to the amount liberated from freshwater sediments and marshes.

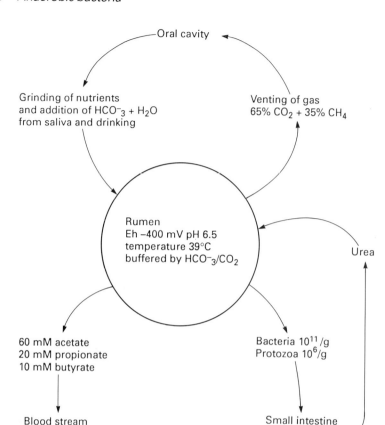

Fig. 4.3 The rumen as a semicontinuous fermentation system.

The digesta pass from the rumen into the small intestine, via the omasum and abomasum. Following lysis by the acid in the abomasum, digestion of the microbial mass that is passed out from the rumen provides the animal with its major source of amino acids and water-soluble vitamins. Urea produced in the large intestine is returned to the rumen via saliva and by diffusion from the bloodstream, providing a source of nitrogen for the rumen microflora. This has two benefits for the ruminant. Firstly, it allows the animal to survive on a diet low in nitrogen and secondly it conserves water which would otherwise be excreted as urine.

The microflora of the rumen is extremely complex but has been studied extensively and its physiological role is well understood. The importance of this flora to the ruminant host can be appreciated when it is considered that the rumen contents make up 10–15% of the animal's total body weight and the microbial flora represents approximately 1% of the dry weight of the entire animal.

The rumen microflora is composed of about 10^{11} bacteria/ml and 10^6 protozoa/ml, with lesser numbers of fungi. The rumen protozoa are obligate anaerobes, chiefly holotrichs and entodiniomorphs. They produce about half of the total fermentation products in the rumen, thus their significance cannot be underestimated. However, they do not appear to be essential; in animals which are defaunated using specific anti-protozoal drugs, the microbial flora continues the rumen fermentation, albeit with a slightly lower efficiency than in untreated animals. Until recently, the existence of obligately anaerobic fungi was doubted, but several species have now been described from the rumen.

The overall rumen fermentation involves the hydrolysis of cellulose and other complex polysaccharides from plant-cell walls, the fermentation of sugars with the production of volatile fatty acids, and the production of methane. A similar fermentation occurs in anaerobic digesters (see Chapter 7). The fermentation of sugars is almost complete, so that ruminants absorb very little glucose; thus they normally have very low blood glucose levels.

The bacterial flora of the rumen is dominated by cellulolytic and methanogenic organisms. The cellulolytic bacteria include *Wolinella succinogenes*, ruminococci and *Butyrivibrio* spp. Most of the bacteria in the rumen are found attached to plant fibres, but clumps of organisms are found in rumen fluid, bound together by extracellular polysaccharides. In the rumen, in contrast to other methanogenic habitats, methane is primarily derived from CO_2 and H_2 since acetate is absorbed by the ruminant host. The most important function of methanogens in the rumen is thus the removal of hydrogen, facilitating the growth of acetogenic bacteria. This phenomenon is known as 'interspecies hydrogen transfer' (see Chapters 5 and 7). The most numerous rumen methanogens are of the genera *Methanobrevibacter*, although *Methanosarcina*, *Methanobacterium* and *Methanomicrobium* also occur.

The activities of the rumen microflora (the standard rumen fermentation) may be summarized thus:

$$100 \text{ [glucose]} \longrightarrow 107 \text{ [acetate]} + 37 \text{ [propionate]} + 28 \text{ [butyrate]} + 59 \text{ [methane]} + \text{heat}$$

The net efficiency of the fermentation is 75%; 18% of the energy is lost as methane and 7% as heat. The average yield of ATP is 4 ATP/mole glucose but this varies slightly depending upon the organism and pathway considered. The efficiency of the fermentation may be increased by the administration of ionophores (such as monensin), which inhibit methanogenesis and stimulate the production of propionic acid. More energy is thus available to the ruminant. The manipulation of rumen metabolism is one area of active research, using growth-promoting antibiotics.

The rumen microflora may also have an adverse effect on the health of the animal. One condition resulting from the rumen metabolism is fog fever, in which pulmonary oedema and emphysema result from the absorption of skatole (3-methylindole). Skatole is produced from tryptophan by lactobacilli in the rumen. A second example of the potentially harmful effects of the rumen microflora is the development of bloat. This condition occurs when gases in the rumen become trapped in a stable foam, preventing their expulsion by eructation. The rumen swells and death may follow compression of other organs. The abdominal wall may be pierced to release the gas; prevention of this condition entails the avoidance of certain types of fodder.

Hindgut fermentation

Other animals, including insects, fish, reptiles, birds and a considerable number of mammals, have evolved a hindgut fermentation. In general these fermentations are not as well understood as that occurring in ruminants. However, in some animals the process is analogous to that in the rumen. Wood-eating insects, such as termites, carry endosymbiotic protozoa in their large intestine. These cellulolytic protozoa produce volatile acids that are absorbed by the insect host.

In gallinaceous birds, such as pheasants and grouse, there are two caeca, located at the junction of the small and large intestines, which together account for about 50% of total body weight. It is probable that acetate, propionate and butyrate produced in the caeca by anaerobes form a substantial proportion of the energy required by the birds.

Rodents and lagomorphs also appear to derive a considerable amount of energy from caecal fermentations. In the rabbit, this may be as much as 12% of the animal's daily energy requirement. The numerically dominant organisms are obligate anaerobes of the genera *Bacteroides*, *Eubacterium*, *Fusobacterium* and *Succinimonas*.

Large herbivorous mammals, for example horses and zebras, have large caeca which may have a volume of up to 30 litres. The colon in these animals has a volume of 50–70 litres. A cellulolytic flora is found in the caecum, of which anaerobes are the major components. The numbers of anaerobes may reach 10^{11}/g caecal contents. The largest populations are of *Bacteroides*, *Butyrivibrio* and *Fusobacterium*. Peptostreptococci and selenomonads are present in lesser numbers.

In omnivorous large mammals, including both pigs and humans, the caecum is not large (Fig. 4.2) and cellulose does not supply a major part of the animal's requirement for energy. A great deal of effort has been expended on the study of the flora of human faeces, based upon the assumption that the faecal flora is representative of the colonic flora. This is true to a certain extent, but does not take into account the mucosal flora of the large intestine, which contains many morphological types not represented in faeces.

In the human gastrointestinal tract the oesophagus and stomach are normally sterile. The small intestine becomes increasingly heavily colonized along its length, so that in the distal portion of the ileum the flora resembles that of the large intestine. Obligate anaerobes are found in the ileum at populations between 10^6 and 10^{10}/g, whereas in the ascending colon, the proximal end of the large intestine, the numbers of anaerobes reach 10^{11}/g. This population density is maintained throughout the large bowel.

The adult human colonic microflora is extremely complex; as many as 300 bacterial species may be present, many of which have not been described in full. Obligate anaerobes comprise 99% or more of the total bacterial numbers in the large bowel. *Bacteroides* spp. are the dominant organisms, comprising up to 30% of the total numbers, of which *B. vulgatus*, *B. thetaiotaomicron* and *B. distasonis* are the most significant species. *Eubacterium* spp. make up 15–20%, and Gram-positive anaerobic cocci such as *Ruminococcus* up to 10%, of the total bacterial numbers. Fusobacteria and bifidobacteria each represent about 5% of the flora. Clostridia are only minor components of the colonic microflora and their numbers rarely exceed 10^6/g in healthy adults. A number of *Clostridium* spp. are however enteropathogens and are discussed in Chapter 6.

The large intestinal flora in the human adult is remarkably stable. Slight differences in flora are observed in different populations. The major factor affecting the composition of the flora is diet. However, minor changes in diet affect the colonic flora in individuals very little and at most have a transient effect. Stress also has a minor effect upon the microflora. However, antibiotic administration can have a very marked adverse effect upon the colonic microflora, with the development of antibiotic-associated diarrhoea (see Chapter 6).

The colonic microflora is far less complex in human neonates. At birth the gastrointestinal tract is sterile, but rapidly becomes colonized from the surroundings and by physical contact with the mother. In breast-fed infants the flora is composed largely of *Bifidobacterium* spp., which prevent colonization by enteropathogenic organisms. In babies fed artificial diets the flora is less homogeneous and putrefactive organisms such as clostridia and *Bacteroides* are of greater importance. In both breast- and bottle-fed infants, the flora gradually changes after weaning to resemble more closely that of the adult.

Until recently, methanogens were not thought to occur in the human colonic microflora. However, the use of appropriate culture techniques (see Chapter 2) has demonstrated that methanogens may be found in the large intestine of 30-70% of adults. SRB are found only in the absence of methanogens, and these two groups are mutually exclusive. Numbers of SRB may be very high (up to 10^{10}/g faeces); *Desulfovibrio* spp. are the most numerous but *Desulfobacter*, *Desulfobulbus*, and *Desulfotomaculum* also occur. SRB appear to out-compete methanogens for hydrogen; sulphated polysaccharides such as mucin are a source of sulphate in the large intestine. The factors which allow methanogens to exclude SRB in some individuals are as yet unknown.

The importance of the bacterial gut flora in humans appears not to lie in its fermentative activity. However, some absorption of volatile acids from the colon does occur (Cummings, 1981). Much of the research on the gut flora in humans has been directed at the potential role of gut bacteria in the aetiology of cancer, particularly of the large bowel. The high incidence of cancer of the colon in North America and western Europe is related to the consumption of a diet high in meat and animal fats. Direct-acting carcinogens are not found in the diet, but several precursors of carcinogens have been identified. One example is 2-amino-3-methyl-3*H*-imidazo [4, 5-*f*]quinoline (IQ). This compound is produced by pyrolysis of amino acids during cooking, particularly in red meats. It is mutagenic after activation by microsomal enzymes and causes splenic and hepatic tumours. However, in the large intestine, IQ is metabolized by *Eubacterium* spp. to the direct-acting mutagen hydroxy-IQ (Bashir *et al.*, 1987).

Other potentially carcinogenic activities of gut bacteria are the transform-ation of steroids and bile acids by *Clostridium* spp. and *Eubacterium* spp., although no mutagenic activity has been detected in any of the resulting compounds. The potential for the formation of carcinogenic nitrosamines by the intestinal flora has also been investigated.

A further significant function of the human intestinal flora is the prevention of colonization by enteropathogenic organisms. This phenom-enon, known as 'colonization resistance', first appears in infants. It appears to depend upon several interacting factors. These include Eh, the presence of volatile acids, competition for nutrients and physical exclusion from niches within the intestinal ecosystem. The fundamental importance of the intestinal flora in this respect is indicated by the effect that antibiotic therapy may have on colonization resistance.

Other body sites

The remaining major site of anaerobic colonization is the female genital tract. The vagina supports a flora composed primarily of Gram-positive anaerobic cocci, *Veillonella*, *Bifidobacterium*, *Lactobacillus* and some *Bacteroides* spp. These organisms ferment available carbohydrates and as a result the healthy vagina normally has a pH not higher than 4. The low pH serves as a barrier to infection by sexually transmitted pathogens such as *Neisseria gonorrhoeae*.

The skin is also colonized by anaerobes, principally *Propionibacterium* spp., the habitat of which is the sebaceous ducts. Anaerobes derived from the intestinal flora can often be isolated from the skin on the back and lower limbs, but this represents contamination or transient colonization.

Soil and plant tissues

Soil is composed of a mixture of inorganic and organic components. There are three basic soil types – clay, sand and silt. Between the particles within soil there are air spaces which may comprise up to one-third of the volume of the soil. Soil particles are generally coated with a fine film of water, which ensures that microbial activity may occur. Organic material reaches the soil from crop residues, leaf litter and from plant roots (via rhizodeposition).

Because of the heterogeneous structure of soil there are numerous micro-habitats within a small space. Thus microbial activity is not uniformly distributed throughout soil. Anaerobic micro-environments arise following microbial metabolism around plant roots and in areas where the water film reduces the oxygen tension on the surface of soil particles. Within the soil population, bacteria are the most numerous organisms, reaching counts of 10^6–10^9 viable cells/g, but they are not the major component of soil biomass, due to their small size. Fungi are far more important in this respect, comprising up to 50% of soil biomass.

Anaerobes are not therefore of overriding importance in soil, but several species of clostridia perform non-symbiotic nitrogen fixation. When anoxic conditions prevail, nitrate reduction poises the Eh at about $+200$ mV until the supply of nitrate is exhausted, after which the Eh falls rapidly to as low as -200 mV and anaerobic nitrogen fixation occurs. The most numerous nitrogen-fixing anaerobes are *C. pasteurianum* and *C. beijerinckii/butyricum*. Numbers of clostridia vary between 10^4/g in open soil and 10^6/g in the rhizosphere. Cellulolytic and pectinolytic clostridia are important in the degradation of plant materials. In heavily waterlogged soils, conditions may become sufficiently reduced for growth of methanogens and SRB to occur. The effects of SRB on metallic structures in soil are discussed in Chapter 7.

Soil anaerobes are universally distributed on root crops and other plant materials. The use of anaerobes in such industrial processes as flax retting relies on the presence of pectinolytic clostridia on the plant stems. However, anaerobes may also be involved in spoilage of root crops and as a cause of food-poisoning (see Chapter 7).

Wetwood

Wetwood is a condition that affects the heartwood of living trees, including both hardwoods and conifers. The heartwood of affected trees is devoid of living parenchyma cells but contains waterlogged xylem tissue. Wetwood results in a significant loss in the value of timber produced from affected trees.

Wetwood is often associated with the production of a malodorous fluid that drains from holes bored into the tree. In addition, methane is produced and may be burnt off as it diffuses from the same boreholes. The fetid odour of the fluid produced results from the presence of several metabolites produced by obligate anaerobes, including acetic, propionic, isobutyric and butyric acids, ethanol, isopropanol and methane.

High numbers of obligate anaerobes may be recovered from the affected tissues (Schink *et al.*, 1981). These may be of three main groups – heterotrophs (10^6/g), nitrogen fixers (10^5/g) and methanogens (10^4/g). Among the organisms frequently isolated are pectinolytic clostridia, *Bacteroides* spp. and *Lactobacillus* spp., as well as a variety of facultatively anaerobic Gram-negative bacilli.

Aquatic environments

Anaerobes perform important functions in the carbon and sulphur cycles in aquatic sediments. Anaerobic activities in aquatic environments differ from aerobic ones because they are not performed by single organisms but result from the interaction of several types of anaerobe. Moreover, in anaerobic processes energy is often transferred to reduced inorganic compounds such as hydrogen sulphide. These compounds are then oxidized by aerobes, so the complete cycling of energy results from the combined action of both aerobes and anaerobes. Methane produced by methanogenic bacteria is returned by methane-oxidizing bacteria to carbon dioxide, which is then converted to biomass by photosynthetic microorganisms, some of which are anaerobic (see Chapter 3). The degree of sulphate reduction occurring depends upon the nature of the aquatic habitat. In marine environments, sulphate is present at a concentration of 29 mmol/litre and thus sulphate

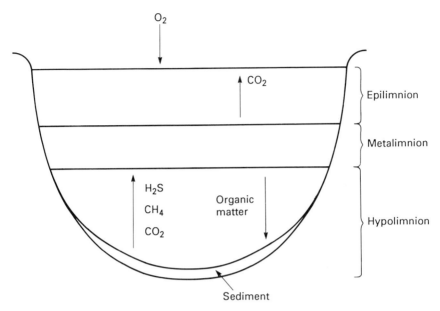

Fig. 4.4 Stratification of a freshwater lake.

reduction is of greater significance than it is in most freshwater environments.

Freshwater habitats

Many freshwater lakes are rich in organic material and are said to be eutrophic. Deep lakes in temperate climates become stratified during the spring because the upper layers of the lake warm up, whereas the deeper water remains at 4°C. This cold layer is known as the 'hypolimnion' and the upper, warmer layer is the 'epilimnion' (Fig. 4.4). The transitional area between hypolimnion and epilimnion is known as the 'thermocline' or 'metalimnion'. Mixing of the upper and lower layers occurs only during spring and autumn, when the whole body of water is at 4°C. The net result is that oxygen is severely limited in the hypolimnion during the summer, because photosynthesis does not take place in the aphotic zone, which usually corresponds to the hypolimnion (light normally penetrates no more than 15 m under the surface).

The aerobic epilimnion is thus the site of oxygenic photosynthesis. Organic material produced in the epilimnion eventually settles to the bottom of the lake, where conditions are poorly oxygenated. Profoundly anaerobic conditions are found a very small distance below the surface of the sediment.

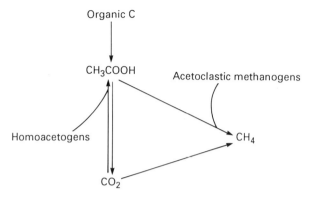

Fig. 4.5 Terminal metabolism of organic carbon in anaerobic freshwater sediments.

Within 1–2 cm the Eh may fall from +200 mV to −200 mV. A range of anaerobic organisms becomes stratified within such an Eh gradient.

Organic material is degraded by hydrolytic and fermentative anaerobes throughout the stratified sediment. At the lowest Eh levels (−200 mV and below) methanogens such as *Methanobacterium*, *Methanobrevibacter*, *Methanogenium*, *Methanospirillum*, *Methanosarcina* and *Methanothrix* occur. Acetate is the major substrate for methanogenesis and as a result interspecies hydrogen transfer is of great importance (see Chapters 5 and 7). Carbon dioxide may also serve as a substrate for methanogenesis, but CO_2 is also metabolized to acetate by homoacetogens (Fig. 4.5). The rate of production of methane may be significant; in one highly eutrophic lake, methane was collected at a rate of 2 ml/min/m^2 sediment (or almost 3 litres/m^2/day).

As observed in the section on the hindgut fermentation, methanogens and SRB are mutually exclusive. In sediments this antagonism is resolved by spatial separation. SRB are found in the sediment above the level at which methanogens are present, where the Eh is about −100 mV, higher than that required by methanogens. In this zone of moderately anaerobic conditions, SRB are the major consumers of hydrogen, due to their more thermo-dynamically efficient use of the gas.

An important factor in the microbiology of sediments is the occurrence of anaerobic methane oxidation. This appears to be significant in controlling the emission of methane from anaerobic sediments. Both SRB and methanogens are able to reduce small amounts of methane (less than 1% of the volume produced). However, further anaerobic methane oxidation may be related to sulphate reduction, involving a consortium of SRB and an as yet unknown anaerobic oxidizer of methane. Aerobic methane-oxidizing bacteria occur at the thermocline and remove the great majority of methane evolved, producing organic carbon and CO_2, which is utilized by oxygenic photo-synthesizers in the epilimnion.

Thus, except in very shallow lakes that are rich in organic material, methane rarely reaches the surface. In paddy fields and marshes, methane oxidizers are not present in significant numbers and vast quantities of methane are released to the atmosphere (estimated to be 2×10^8 tonnes per annum globally).

As H_2S diffuses upwards through the sediment it may be oxidized by aerobes (*Thiobacillus*) or microaerophiles (*Beggiatoa*). In shallow lakes (less than 10 m in depth) the upper layers of sediment may be a suitable habitat for anoxygenic phototrophs such as the purple sulphur bacteria (Chromatiaceae) and the green sulphur bacteria (Chlorobiaceae). When both types of phototroph are present they become stratified by Eh, the green sulphur bacteria growing deeper in the sediment than the purple sulphur bacteria. Both types of anoxygenic phototrophs deposit sulphur, the green photobacteria doing so extracellularly and the purple bacteria intracellularly. Heavy growth of these bacteria may result in blooms of various colours, depending on the types of phototrophs involved. In such an environment, sulphate may become exhausted; methanogens will then become dominant. This occurs particularly in small ponds, especially those containing excess organic matter (such as leaves).

Permanent associations of sulphate reducers and sulphide oxidizers are known as 'sulphureta'. Sulphureta are examples of anaerobic primary-producing systems. They are found in salt marsh sediments, cyanobacterial mats, sulphur springs, marine sediments and other habitats. The deposition of sulphur by sulphureta may be sufficient to allow commercial production of sulphur from these sources. It is probable that enormous sulphureta once existed which were responsible for the present underground deposits of sulphur.

A variety of anaerobes may be found in hot springs, including SRB, methanogens such as *Methanobacterium*, *Methanococcus*, *Methanogenium* and *Methanothermus*, and the acidophilic, Archaebacterial thermophile *Thermoproteus*.

Anaerobic degradation of organic material also occurs in other extreme environments such as hypersaline lakes. Halophilic anaerobes isolated to date have all been fermentative chemoorganotrophs (Oren, 1986). Sulphate reduction, presumably by bacteria, has been demonstrated in several hypersaline environments, but SRB have not been isolated. The high sulphate concentrations in hypersaline environments ensure that methanogenesis cannot proceed using hydrogen or acetate as substrates because of the more efficient scavenging by SRB. Thus methanogenic substrates that are not utilized by SRB are of more importance, including trimethylamine, methanol and methionine. A number of halophilic methanogens have been isolated, tolerating salt concentrations up to 3.5 M.

Marine environments

In marine environments, the concentration of dissolved organic matter is much lower than in freshwater (<1 mg/litre). In general, the wave motion of oceans ensures that the water column is oxygenated down to great depths. Anaerobic conditions and stratification occur in marine sediments, but SRB are the dominant organisms involved in anaerobic degradation, due to the uniformly high concentration of sulphate. Methanogenesis also occurs; methanogenic activity has been recorded in cores as deep as 1000 m below the sediment surface.

A variety of unusual organisms have been recovered from submarine hydrothermal vents. In these habitats mineral-laden water at 350°C mixes with the seawater. Thermophilic methanogens (*Methanococcus*) and novel barotolerant Archaebacteria have been isolated. The disc-shaped genera *Thermodiscus* and *Pyrodictium* have temperature optima of 87°C and 105°C, respectively. Both genera metabolize sulphur.

References and further reading

Bashir, M., Kingston, D. G. I., Carman, R. J., van Tassell, R. L. and Wilkins, T. D. (1987) Anaerobic metabolism of 2-amino-3-methyl-3*H*-imidazo[4, 5-*f*]quinoline (IQ) by human fecal flora. *Mutation Research* **190**, 187–190.

Clarke, R. T. J. and Bauchop, T. (1977) *Microbial Ecology of the Gut*. London: Academic Press.

Cummings, J. H. (1981) Short chain fatty acids in the human colon. *Gut* **22**, 763–779.

Drasar, B. S. and Hill, M. J. (1974) *Human Intestinal Flora*. London: Academic Press.

Gibbons, R. J. and van Houte, J. (1980) Bacterial adherence and the formation of dental plaques. In: *Bacterial Adherence* (ed. Beachey, E. H.), pp. 63–104. London: Chapman and Hall.

Grimble, G. (1989) Fibre, fermentation, flora, and flatus. *Gut* **30**, 6–13.

Large, P. J. (1983) *Methylotrophy and Methanogenesis*. Wokingham: Van Nostrand Reinhold.

Lee, A. (1985) Neglected niches. The microbial ecology of the gastrointestinal tract. In: *Advances in Microbial Ecology*, Vol. 8 (ed. Marshall, K. C.), pp. 115–162. New York: Plenum Press.

Macy, J. M. and Probst, I. (1979) The biology of the gastrointestinal *Bacteroides*. *Annual Review of Microbiology* **33**, 561–594.

Nedwell, D. B. (1984) The input and mineralization of organic carbon in anaerobic aquatic sediments. In: *Advances in Microbial Ecology*, Vol. 7 (ed. Marshall, K. C.), pp. 93–131. New York: Plenum Press.

Oren, A. (1986) The ecology and taxonomy of anaerobic halophilic eubacteria. *FEMS Microbiology Reviews* **39**, 23–29.

Oren, A. (1988) Anaerobic degradation of organic compounds at high salt concentrations. *Antonie van Leeumenhoek* **54**, 267–277.

Postgate, J. R. (1984) *The Sulphate-reducing Bacteria*, 2nd edn. Cambridge: Cambridge University Press.

Schink, B., Ward, J. C. and Zeikus, J. G. (1981) Microbiology of wetwood: role of anaerobic bacterial populations in living trees. *Journal of General Microbiology* **123**, 313–322.

Wolin, M. J. (1979) The rumen fermentation: a model for microbial interactions in anaerobic ecosystems. In: *Advances in Microbial Ecology*, Vol. 3 (ed. Alexander, M.), pp. 49–77. New York: Plenum Press.

5

Biochemistry and metabolism of anaerobes

Anaerobes and anaerobiosis

The likely evolution of present-day anaerobes was discussed briefly in Chapter 1. It was seen that anaerobic bacteria are of extremely ancient origin, and that they are found in both the major taxa of prokaryotes, the Archaebacteria and the Eubacteria. It should not therefore surprise us that not all anaerobes are affected by oxygen in the same way. However, for many years a unifying theory of anaerobiosis was sought, with little success. In order to understand the behaviour of anaerobes in relation to air it is first necessary to review some of the more plausible hypotheses made to explain the sensitivity of obligate anaerobes to oxygen. These include:

 (i) bimolecular oxygen is directly toxic to the cell;
 (ii) oxygen raises the Eh above a limiting level for the growth of a particular species;
 (iii) sulphydryl groups within key enzymes are oxidized by oxygen thus preventing growth;
 (iv) in the presence of oxygen the cell diverts its reducing activity to the reduction of oxygen rather than to its metabolic activities;
 (v) the breakdown products of oxygen, particularly the free radicals, are toxic to cellular components.

Direct toxic action of oxygen

It is difficult to design experiments to measure in isolation the effect of any one of the above hypotheses. However, damage seems to occur when oxygen is actually consumed by cells, implying that molecular oxygen itself is not toxic to anaerobes.

Effect of oxygen on Eh

Anaerobes develop and maintain a low redox potential (Eh) in their growth media. The difficulties associated with the measurement and interpretation of the Eh of bacterial cultures were discussed by Morris (1975). Briefly, the Eh of a bacterial culture is the result of numerous interactions between redox couples. Eh is also affected by the medium used and the size of the inoculum. It is extremely difficult to specify exactly what the measured Eh of a culture actually means, but it remains a useful indicator in the absence of any more relevant marker of anaerobiosis. Growth of obligate anaerobes is only possible below a certain Eh. Generally, the more oxygen-sensitive anaerobes require a lower Eh before growth will be initiated. Thus the raising of Eh above this limiting level by oxygen is an attractive explanation for inhibition of anaerobic growth under aerobic conditions. The protective effects of reducing agents also support this hypothesis. However, it has repeatedly been shown that several clostridial species will grow in media with an Eh of >300 mV if the Eh is maintained by ferricyanide, but will not grow in the same media when the Eh is raised to the same level by aeration. Thus, it is clear that raising of Eh by oxygen can at best only partly explain the inhibitory effects of oxygen upon obligate anaerobes.

Inhibition of sulpfhydryl groups by oxygen

Although anaerobes contain many oxygen-sensitive enzymes, of which sulpfhydryl groups are important components, there is little or no evidence that oxygen inhibits these or any other specific cellular constituents directly.

Diversion of reducing activity

Evidence for this hypothesis derives from experiments which show that oxygen oxidizes NADPH at a rate greater than that at which it is produced by fermentation. Thus, electrons are drained from intracellular electron donors and growth ceases. In several of the butyric clostridia, oxygenation of cultures results in a diminution of the amount of butyrate produced, with a concomitant increase in acetate and pyruvate excretion. Butyrate production is resumed immediately cultures are rendered anaerobic. This hypothesis therefore remains a plausible explanation for the inhibition of anaerobic growth by oxygen.

Breakdown products of oxygen

Highly reactive, potentially destructive products of oxygen reduction are invariably formed whenever oxygen is consumed by living cells and has the opportunity to react with reduced cellular components (such as thiols, iron-sulphur proteins, flavoproteins and tetrahydropteridines):

$$O_2 + e^- \longrightarrow O_2^-$$
superoxide anion

$$O_2 + 2e^- \longrightarrow H_2O_2$$
hydrogen peroxide

$$O_2 + 3e^- \longrightarrow H_2O + OH^-$$
hydroxyl radical

The superoxide anion is a highly reactive free radical. It is the longest lived of the free radicals derived from oxygen and can serve to initiate free-radical chain reactions, in addition to being a powerful oxidizing agent. Superoxide anions may also be formed within cells by leaking of electrons from ubiquinones in electron-transport chains. Within the cell, superoxide anions react to form hydrogen peroxide. This is known as the 'dismutation reaction':

$$2O_2^- + 2H^+ \longrightarrow H_2O_2 + O_2$$

Hydrogen peroxide is itself damaging to cells, but further superoxide anions react with hydrogen peroxide in the Haber–Weiss reaction to form hydroxyl radicals:

$$O_2^- + H_2O_2 + H^+ \longrightarrow O_2 + H_2O + OH^-$$

The Haber–Weiss reaction is catalysed *in vivo* by trace amounts of iron–copper complexes. Hydroxyl radicals react very quickly with almost all components of cells, including DNA, proteins and carbohydrates. Removal of hydrogen atoms from membrane lipids initiates lipid peroxidation.

A further reaction yields singlet oxygen:

$$O_2^- + OH^- \longrightarrow OH^- + {}^1O_2$$

These products attack lipid cell membranes, cause breaks in DNA and inhibit many enzymes in anaerobes (such as hydrogenase, nitrogenase and pyruvate dehydrogenase).

Defensive mechanisms against oxygen toxicity

Every organism which uses oxygen must possess mechanisms which protect it against these toxic derivatives of oxygen. Thus, aerobes and facultative anaerobes have evolved a number of strategies for their defence against the deleterious effects of oxygen and its products. There are four approaches which are available to cells:

(i) removal of superoxide anions;
(ii) removal of hydrogen peroxide;
(iii) reduction of levels of iron–copper complexes to a minimum;
(iv) rendering of membranes refractory to lipid peroxidation.

All these approaches are adopted by microorganisms. Many obligate aerobes (such as *Pseudomonas* and *Neisseria*) possess cytochrome oxidases, which add electrons singly to oxygen, forming superoxide anions. However, the radicals are strongly bound to the reaction site until a total of four electrons have been added and water is then released. This contrasts with the leaking of superoxide radicals from ubiquinones.

The removal of superoxide anions is accomplished by superoxide dismutases (SOD), metalloprotein enzymes which catalyse the dismutation reaction:

$$O_2^- + O_2^- + 2H^+ \longrightarrow H_2O_2 + O_2$$

The uncatalysed dismutation reaction occurs at a significant rate, but in the presence of SOD the reaction proceeds much faster, and over a wide range of pH values.

Catalase and peroxidases remove hydrogen peroxide. Catalase breaks down hydrogen peroxide and thus prevents formation of hydroxyl radicals:

$$2H_2O_2 \longrightarrow 2H_2O + O_2$$

Theories of anaerobiosis

On the basis of the observations made above it is possible to combine some of the previous hypotheses concerning anaerobiosis into a biphasic explanation of oxygen toxicity, applicable to most, if not all, obligate anaerobes. For most moderate anaerobes (such as *Bacteroides* or *Clostridium*) there are two phases of oxygen toxicity. The first of these phases is a bacteriostatic effect and is due directly to the effect of oxygen. It occurs when an anaerobic organism is exposed to oxygen at a greater concentration than it can tolerate while still growing. This Phase 1 effect is reversible and the organism will recommence growth when the Eh is reduced again to a sufficiently low level.

The organism defends itself against Phase 1 oxygen toxicity by diverting its reductive energy into reducing the oxygen rather than into metabolic processes, and thus growth ceases. If the reductive Phase 1 defences of a moderate anaerobe are overcome, then a bactericidal Phase 2 of oxygen toxicity results.

It was once thought that anaerobiosis was simply the result of a lack of catalase. However, some obligate anaerobes possess catalase (such as *Bacteroides fragilis*). This hypothesis was succeeded by one which held that SOD was lacking in obligate anaerobes. This enzyme is also found in a wide variety of anerobes including some methanogens. To be well equipped to withstand exposure to oxygen, an organism must contain catalase, SOD and quenchers of molecular oxygen (such as carotenoids). The most oxygen-sensitive anaerobes, such as methanogens, do not generally possess SOD or catalase. The great variation in extent of oxygen sensitivity between species and strains of anaerobes may be at least in part explained by the presence or absence of protective enzymes. As an example, recent work on black-pigmented *Bacteroides* spp. has suggested that a combination of inducible NADH-oxidases and SOD may be responsible for aerotolerance (Amano *et al.*, 1988). However, exceptions to this rule do exist and there is no unitary theory of anaerobiosis.

Energy-yielding metabolism in anaerobes

The continuous requirement of living cells for energy may be met from chemical or physical sources. Within cells, energy is carried by adenosine 5′-triphosphate (ATP). Intracellular energy-yielding processes are coupled to the endergonic production of ATP from adenosine 5′-diphosphate (ADP) and phosphate. ATP is then used as a source of energy for other work within the cell. The free energy of hydrolysis of ATP to ADP plus phosphate is $\Delta G^{o\prime} = -31.8$ kJ/mole and that of ATP to adenosine 5′-monophosphate (AMP) plus pyrophosphate is $\Delta G^{o\prime} = -41.7$ kJ/mole. Thus, ATP is an important donor of phosphoryl groups in enzyme-mediated reactions.

It has been calculated that within anaerobic bacteria, between 41.8 and 50.2 kJ is required for the synthesis of 1 mole of ATP, depending upon intracellular pH (Thauer *et al.*, 1977). Thus, a catabolic reaction must generate a free-energy change of at least -41.8 to -50.2 kJ in order to facilitate the production of 1 mole of ATP.

Energy-yielding reactions and ATP formation are coupled stoichiometrically, so that energy transformation occurs in 'packets' of 42–50 kJ/mole. However, it is important to recognize that neither anabolic nor catabolic processes within the cell proceed with 100% efficiency. A variable proportion of the transformed energy is dissipated as heat.

Efficiencies of more than 80% have not been recorded. The great majority of anaerobes work at much lower efficiencies (25–50%). In aerobic respiration the thermodynamic efficiency is not markedly higher but the yield of ATP is much greater. Fermentation of glucose yields 2 moles of ATP per mole of substrate, whereas aerobic respiration yields 38 moles of ATP per mole of substrate.

The energy necessary for ATP synthesis is almost invariably generated by redox reactions. Redox (or oxidation–reduction) reactions occur between pairs of compounds, which exist in two states, one compound being oxidized while the other is reduced, or vice versa. The redox potential of the couple is usually expressed as the value of E_o' (see Chapter 2). By comparing E_o' values it is possible to determine whether one couple will be oxidized or reduced by another couple. The amount of free energy released when two couples undergo a redox reaction ($\Delta G^{o'}$) is dependent upon the difference between the values of E_o' for the two couples ($\Delta E_o'$).

The high yield of energy from oxidative phosphorylation is possible because of the $\Delta E_o'$ between the couples NADH/H$^+$ ($E_o' = -320\,\mathrm{mV}$) and O_2/H_2O ($E_o' = +810\,\mathrm{mV}$); $\Delta E_o' = 1130\,\mathrm{mV}$ and $\Delta G^{o'} = -217\,\mathrm{kJ/mole}$. Clearly anaerobes, which cannot use oxygen as the terminal electron acceptor, must use other, less-efficient substances in this role.

Because of the inefficiency of anaerobic energy metabolism, anaerobic processes are associated with the formation of large amounts of reduced compounds which are excreted. Most of these are organic compounds such as ethanol, butyrate and methanol, but H_2 and CO_2 are produced, and sulphate-reducing bacteria (SRB) elaborate H_2S.

There are three ways in which anaerobes may gain energy:

(i) substrate-level phosphorylation (SLP) or fermentation;
(ii) electron-transport phosphorylation (ETP) or anaerobic respiration;
(iii) photophosphorylation.

Fermentation

Fermentation is an energy-yielding process in which ATP is generated by SLP. The terminal electron acceptor is an intermediate metabolite and both it, and the electron donor, are organic compounds. An example is the lactate fermentation. In this fermentation, pyruvate serves as the terminal electron acceptor, which is reduced to lactate.

One characteristic of anaerobic metabolism is that many catabolic pathways are branched (Fig. 5.1), the branches having differing thermodynamic efficiencies and yields of ATP per mole of substrate. The relative rates of the branches are then adjusted to optimize the overall thermodynamic efficiency and ATP yield for the environment surrounding the organism. One effect of this variability is reflected in the ratio of fermentation end-products elaborated under different conditions. An example of this type of branched pathway is the fermentation of glucose by many clostridia,

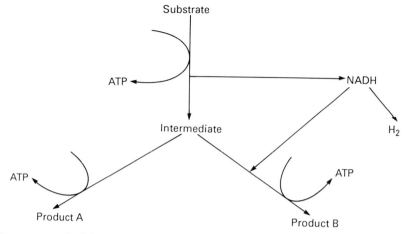

Fig. 5.1 Branched fermentation pathway with variable thermodynamic efficiency and ATP production.

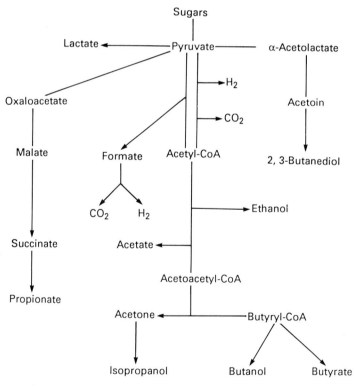

Fig. 5.2 Production of fermentation end-products via pyruvate. Steps leading to production of products are indicated by arrows.

Table 5.1 Some 'energy-rich' compounds involved in SLP reactions

Type of compound	Energy-rich compound	$G^{o\prime}$ of hydrolysis (kJ/mole)*
Acyl thioester	Acetyl CoA	-35.7
	Propionyl CoA	-35.6
	Butyryl CoA	-35.6
	Succinyl CoA	-35.1
Phosphoacyl anhydride	Acetyl phosphate	-44.8
	Butyryl phosphate	-44.8
	Biphosphoglycerate	-51.9
	Carbamyl phosphate	-39.3
Acyl anhydride	N^{10}-formyltetrahydrofolate	-23.4
Phosphoenol-ester	Phosphoenolpyruvate	-51.6

* $G^{o\prime}$ is the free-energy change at pH 7, a free Mg^{2+} concentration of 10^{-3} molar and an ionic strength of 0.25.

which yields variable proportions of acetate and butyrate under different cultural conditions:

$$\text{glucose} + 4H_2O \longrightarrow 2CH_3COO^- + 2HCO_3^- + 4H^+ + 4H_2 + 4ATP$$
$$(\Delta G^{o\prime} = -200 \text{ kJ/mole})$$

$$\text{glucose} + 2H_2O \longrightarrow CH_3CH_2CH_2COO^- + 2HCO_3^- + 3H^+ + 2H_2 + 3ATP$$
$$(\Delta G^{o\prime} = -255 \text{ kJ/mole})$$

Natural butyrate fermentations, such as that by *C. butyricum*, yield approximately 0.6 mole acetate and 0.7 mole butyrate per mole of glucose, with the generation of approximately 3.3 moles of ATP.

Many anaerobes use exclusively SLP for the generation of ATP. A wide variety of substrates are fermented by such organisms, yielding a number of end-products including butyrate, caproate, lactate, acetone, ethanol, isopropanol, *n*-butanol, 2,3-butanediol, CO_2 and H_2 (Fig. 5.2). The butyric fermentation is initiated by the breakdown of sugars in the Embden–Meyerhof–Parnas pathway, yielding pyruvate. This butyric pathway is shared by many species of Gram-negative anaerobes (*Acidaminococcus* and *Fusobacterium*) and many Gram-positive anaerobes (*Clostridium*, *Sarcina*, *Eubacterium* and *Butyrivibrio*). Despite the very wide range of fermentable substrates, there are actually very few 'energy-rich' compounds that can be used to yield energy in SLP reactions (Table 5.1).

Phosphoenolpyruvate is an intermediate in the fermentation of carbohydrates and is an important source of ATP synthesis in saccharolytic anaerobes. Acetyl phosphate is also a very important phosphate donor in SLP reactions. Butyryl phosphate is the intermediate in the butyric fermentation. Carbamyl phosphate is the important energy-rich compound in the fermentation of arginine by *C. perfringens* and *C. botulinum*:

$$\text{arginine} + H_2 \longrightarrow \text{citrulline} + NH_3$$
$$\text{citrulline} + P_i \longrightarrow \text{ornithine} + \text{carbamyl phosphate}$$
$$\text{carbamyl phosphate} + ADP \longrightarrow \text{carbamate} + ATP$$

The use of intermediates as terminal electron acceptors in the fermentation pathway occurs because most fermentations are balanced oxidation–reduction processes. A rather more specialized fermentation occurs in some clostridia (for example *C. sporogenes*). In the Stickland reaction pairs of amino acids are fermented, one serving as the primary electron donor and being oxidized, while the other acts as the terminal electron acceptor and is reduced:

$$CH_3CH(NH_2)COOH + 2H_2O \longrightarrow CH_3COOH + CO_2 + NH_3 + 4H^+$$
alanine (donor)   acetate

$$2CH_2(NH_2)COOH + 4H^+ \longrightarrow 2CH_3COOH + 2NH_3$$
  glycine (acceptor)   acetate

The overall reaction is:

$$CH_3CH(NH_2)COOH + 2CH_2(NH_2)COOH + 2H_2O$$
    alanine   glycine

$$\longrightarrow 3CH_3COOH + 3NH_3 + CO_2$$
          acetate

Commonly fermented donor : acceptor pairs include leucine : proline, iso-leucine : hydroxyproline, valine : ornithine and alanine : glycine. The Stickland reaction allows almost all the amino acid constituents of proteins to be utilized as sources of energy.

Many butyric acid-producing clostridia also ferment single amino acids via pyruvate. However, other species either possess saccharoclastic ability and cannot ferment amino acids, or ferment amino acids but are unable to utilize carbohydrates. Many, but not all, amino acid-fermenting clostridia are actively proteolytic but others are not and therefore depend upon the availability of free amino acids in the growth medium.

A number of clostridia metabolize carbohydrates by other than the butyric acid pathway (Fig. 5.2). These include organisms that ferment cellulose and produce ethanol, acetate, formate, lactate and succinate. One species, *C. kluyveri*, ferments ethanol and acetate, producing butyrate, caproate and H_2.

A unique fermentation is practised by *Bifidobacterium* spp., which

Table 5.2 Redox potentials of some donor:acceptor pairs in ETP

Redox couple	E_o' (mV)
SO_4^{2-}/HSO_3^-	−516
Ferredoxin ox./red. (E_{01}')	−398
NAD/NADH	−320
CO_2/acetate$^-$	−290
CO_2/CH_4	−244
Pyruvate$^-$/lactate$^-$	−190
HSO_3^-/HS^-	−116
Fumarate/succinate	+33
O_2/H_2O	+818

Data from Thauer *et al.* (1977).

produce lactate and acetate from glucose without the formation of pyruvate; acetyl phosphate is the energy-rich intermediate involved in this pathway. This pathway is known as the 'bifid shunt' and is made possible by the production of fructose-6-phosphate phosphoketolase.

Anaerobic respiration

Anaerobic respiration is an energy-yielding process utilizing an electron-transport chain located in the bacterial cytoplasmic membrane, in which oxygen is not the terminal electron acceptor. The electron-transport chain may be similar to those found in aerobes but the terminal electron acceptor can be sulphate, formate, nitrate, nitrite or CO_2. Electron transport-linked phosphorylation (ETP) is thus dependent upon the electrochemical potential between redox partners of different redox potential, which is used to drive the production of ATP. The redox potentials of some donor:acceptor pairs are shown in Table 5.2. The free-energy change associated with ETP in anaerobes is thus much less than that of oxidative phosphorylation (Table 5.3). Accordingly, the growth yields of anaerobes are low and anaerobes often grow more slowly than aerobes.

The reduction of carbon dioxide to methane, dependent upon the presence of dihydrogen, is a capability unique to the methanogenic archaebacteria:

$$CO_2 + 4H_2 \longrightarrow CH_4 + 2H_2O \ (\Delta G^{o'} = -131 \text{ kJ/mole})$$

A number of electron transport proteins have been detected in methanogens, including flavoproteins and cytochromes. However, the major

Table 5.3 Free-energy change associated with ETP using hydrogen as the reductant

Redox couple	$G^{o'}$ (kJ/mole hydrogen)
CO_2/CH_4	-16.4
Sulphate/sulphide	-18.8
Fumarate/succinate	-43.1
Nitrate/nitrite	-81.6
O_2/H_2O	-118.4
Nitrite/nitrogen	-132.6

Data from Morris (1986).

components of the electron-transport pathway in methanogens differ from those in Eubacteria and have yet to be fully characterized. The electron carriers are membrane bound and most of them are very oxygen sensitive. The redox potentials at which most operate are extremely low, including hydrogenase ($E_o' = -414$ mV), coenzyme M ($E_o' = -193$ mV), coenzyme F_{420} ($E_o' = -373$ mV), factor F_{342} ($E_o' = -342$ mV) and factor B_o' ($E_o' = -450$ mV). The structure and function of only two of these compounds, coenzyme M and coenzyme F_{420}, are understood.

Coenzyme F_{420}, a deazaflavin, is an important electron acceptor and acts as a coenzyme for a number of enzymes, which include hydrogenase, formate dehydrogenase, pyruvate dehydrogenase, α-ketoglutarate dehydrogenase, F_{420}-dependent $NADP^+$ reductase and carbon monoxide dehydrogenase. Its function appears similar to that of ferredoxin, which is found in Eubacterial anaerobes but not in methanogens. Coenzyme F_{420} may be involved in methanogenesis in the transfer of electrons from H_2 to factor B_o' (methanopterin). Nickel is essential for the synthesis of hydrogenase, carbon monoxide dehydrogenase and methyl-coenzyme M reductase.

The carrier molecule for the reduction of the methyl group in methanogenesis is known as coenzyme M (2-mercaptoethanesulphonic acid). It is an important component of the methyl-coenzyme M methylreductase system in hydrogenotrophic and acetotrophic methanogens. This complex process involves three components (A, B and C), ATP and Mg^{2+} ions. Component A is a hydrogenase, B is a coenzyme and C is the methylreductase. The methylated coenzyme M serves as the substrate for the reaction:

$$CH_3\text{--}S\text{--}CoM \longrightarrow CH_4 + HS\text{--}CoM$$

While the favoured substrates for growth of methanogens are H_2 and CO_2, most species can also utilize formate, while *Methanosarcina* spp. and

Methanococcus spp. can also metabolize acetate, methanol and methyl-amines. Other methanogens can utilize carbon monoxide. Of these substrates, reduction of acetate is by far the least thermodynamically favourable:

$$4HCOO^- + 4H^+ \longrightarrow CH_4 + 3CO_2 + 2H_2O \; (\Delta G^{o\prime} = -111 \text{ kJ/mole})$$
$$CH_3COO^- + H^+ \longrightarrow CH_4 + CO_2 \; (\Delta G^{o\prime} = -36 \text{ kJ/mole})$$
$$4CH_3OH \longrightarrow 3CH_4 + CO_2 + 2H_2O \; (\Delta G^{o\prime} = -106 \text{ kJ/mole})$$
$$4CH_3NH_2 + 2H_2O \longrightarrow 3CH_4 + CO_2 \; (\Delta G^{o\prime} = -55 \text{ kJ/mole})$$

One very important function of methanogenic organisms is the removal of hydrogen from the environment in which they grow. This facilitates continuing fermentation by acetogenic bacteria and is known as interspecies hydrogen transfer (see Chapters 4 and 7). This phenomenon occurs when the flow of fermentation-generated electrons is transferred away from the production of reduced organic end-products to proton reduction. Production of hydrogen then becomes the major electron sink in the system. Because of the thermodynamic properties of this reaction, the shift in electron flow requires a mechanism for continuous removal of hydrogen as soon as it is formed. Interspecies hydrogen transfer provides mechanisms by which:

(i) relatively unfermentable substrates can be used as sources of carbon and energy in anaerobic environments;
(ii) some saccharolytic bacteria can more completely oxidize their substrates;
(iii) methanogens may obtain CO_2 and H_2 from substrates that they cannot metabolize directly.

In mixed cultures containing methanogens, the partial pressure of hydrogen may be as low as 1.5×10^{-3} atm and this makes the oxidation of NADH thermodynamically favourable.

Sulphate-reducing bacteria (SRB) are responsible for dissimilatory sulphate reduction, utilizing sulphate as the terminal electron acceptor:

$$SO_4^{2-} + 4H_2 + H^+ \longrightarrow HS^- + 4H_2O \; (\Delta G^{o\prime} = -158 \text{ kJ/mole})$$

Other sulphur-containing compounds such as sulphite and thiosulphate and even elemental sulphur may also be metabolized. *Desulfuromonas* spp. cannot utilize sulphate, sulphite or thiosulphate as terminal electron acceptors, but can use elemental sulphur, cystine, oxidized glutathione and fumarate. The electron donors used include lactate, pyruvate, ethanol and formate.

$$CH_3COO^- + SO_4^{2-} + 3H^+ \longrightarrow 2CO_2 + H_2S + 2H_2O$$
$$(Desulfobacter\ spp.:\ \Delta G^{o\prime} = -63\ kJ/mole)$$
$$CH_3COO^- + 4S + 2H_2O + H^+ \longrightarrow 2CO_2 + 4H_2S$$
$$(Desulfuromonas\ spp.:\ \Delta G^{o\prime} = -39\ kJ/mole)$$

Some SRB possess a complete tricarboxylic acid cycle and can oxidize acetate completely to carbon dioxide; others do not, and accumulate acetate as an end-product.

The electron-transport system of SRB is located on the cell membrane and includes dehydrogenases, electron carriers and a number of reductases. The electron carriers of SRB include cytochrome c_3 ($E_o' = -300\ mV$), cytochrome c_{553} ($E_o' = -100\ mV$), flavodoxin ($E_o' = -400\ mV$) and rubredoxin ($E_o' = -60\ mV$), in addition to membrane-bound cytochromes and menaquinone-6 ($E_o' = -74\ mV$).

In SRB that cannot oxidize acetate, a soluble hydrogenase generates dihydrogen, from hydrogen atoms derived from the oxidation of other compounds such as lactate. In the presence of sulphate the dihydrogen is oxidized by a periplasmic hydrogenase with the concurrent production of sulphate. This is known as 'hydrogen cycling' and results in a pumping of electrons (or proton motive force).

Archaebacteria of the family Thermoproteales are sulphur reducers, using dihydrogen (*Pyrodictium* and *Thermoproteus*) and/or organic hydrogen (*Desulfurococcus*, *Thermofilum* and *Thermococcus*) as electron donors:

$$H_2 + S \longrightarrow H_2S\ (\Delta G^{o\prime} = -33.6\ kJ/mole)$$

Many obligate anaerobes and some facultative anaerobes, as well as several species of helminths, use fumarate as an electron acceptor in ETP:

$$[fumarate]^{2-} + 2H_2 \longrightarrow [succinate]^{2-}\ (\Delta G^{o\prime} = -86\ kJ/mole)$$

The obligate anaerobes capable of fumarate reduction with NADH as the electron donor include *Anaerovibrio lipolytica*, several *Bacteroides* spp., *Selenomonas ruminantium* and *Ruminococcus flavefaciens*. Other electron donors used include formate (*Clostridium formicaceticum* and *Wolinella succinogenes*) and lactate (*Desulfovibrio gigas* and *Veillonella alcalescens*). The redox potential of the fumarate:succinate couple is high enough ($E_o' = +33\ mV$; see Table 5.2) to permit fumarate oxidation by a variety of electron donors, which yields sufficient free energy to synthesize 1 mole of ATP. The electron carriers in these organisms are menaquinones and cytochrome *b*. The electron-transport chain in *W. succinogenes* has been extensively studied (Fig. 5.3).

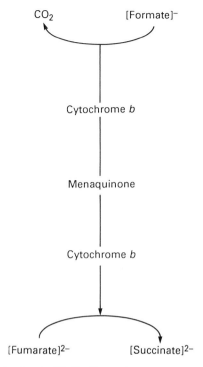

Fig. 5.3 Fumarate reduction in *Wolinella succinogenes*.

Many facultative anaerobes and some obligate anaerobes such as *V. al-calescens*, *W. succinogenes*, *Clostridium perfringens* and *S. ruminantium* can use nitrate as the terminal electron acceptor:

$$NO_3^- + H_2 \longrightarrow NO_2^- + H_2O \ (\Delta G^{\circ\prime} = -163.2 \text{ kJ/mole})$$

The reduction of nitrate to nitrite occurs via a membrane-associated electron-transport system composed of dehydrogenases, electron carriers and nitrate reductase. In obligate anaerobes the electron carriers are menaquinone and cytochrome *b*, with the exception of *C. perfringens*, in which electrons are carried by ferredoxin. The electron donors in anaerobes include NADH, lactate, hydrogen and formate; in *C. perfringens*, pyruvate is the electron donor in nitrate reduction. Nitrate reduction in *C. perfringens* is not linked to phosphorylation.

Photophosphorylation

The photosynthetic apparatus of anaerobic phototrophs contains only one reaction centre, in contrast to the cyanobacteria and green plants which possess an ancillary reaction centre. In the purple bacteria (Rhodospirillales)

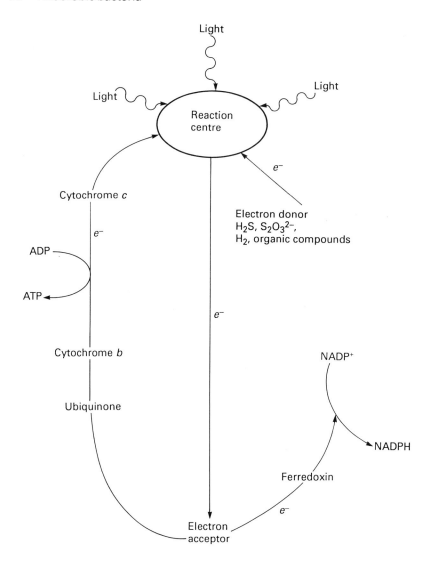

Fig. 5.4 Cyclic photophosphorylation in anoxygenic photosynthesis.

the photosynthetic pigment is either bacteriochlorophyll *a* or *b*, located wholly on the cell membrane. Within the green phototrophs (Chlorobiales) bacteriochlorophyll *c*, *d* or *e* is located within the chlorosomes or 'chlorobium vesicles'. Bacteriochlorophylls absorb light at wavelength maxima of 715–1035 nm.

The green and purple bacteria are anoxygenic, that is they do not generate oxygen during photosynthesis. Because they lack a second reaction centre, these anoxygenic phototrophs require electron donors other than hydroxyl ions, and utilize reduced sulphur compounds, molecular hydrogen or simple organic compounds.

The mechanism of photophosphorylation in anoxygenic phototrophs is cyclic (Fig. 5.4). The primary electron acceptor is either an iron–protein or iron–quinone complex. The electron is then passed via ubiquinone and cytochromes *b* and *c*, back to the bacteriochlorophyll located in the reaction centre.

NADPH is generated via the primary electron acceptor and ferredoxin by ATP-driven reversed electron transport, and is then used to fix carbon dioxide via the Calvin cycle (photolithotrophs) or by the reductive pathway of the tricarboxylic acid cycle (in *Chlorobium* spp.).

In green sulphur bacteria such as *Chlorobium* spp., sulphide is the electron donor and sulphur is deposited extracellularly:

$$CO_2 + 2H_2S \xrightarrow{light} (CH_2O)_n + 2S + H_2O$$

Purple sulphur bacteria are photolithotrophs, since they can oxidize sulphur and sulphide to sulphate.

References and further reading

Amano, A., Tamagawa, H., Takagaki, M., Murakami, Y., Shizukuishi, S. and Tsunemitsu, A. (1988) Relationship between enzyme activities involved in oxygen metabolism and oxygen tolerance in black-pigmented *Bacteroides*. *Journal of Dental Research* **67**, 1196–1197.

Archer, D. B. and Harris, J. E. (1986) Methanogenic bacteria and methane production in various habitats. In: *Anaerobic Bacteria in Habitats Other than Man* (eds Barnes, E. M. and Mead, G. C.), pp. 185–223. Oxford: Blackwell Scientific Publications.

Barker, H. A. (1981) Amino acid degradation by anaerobic bacteria. *Annual Review of Biochemistry* **50**, 23–40.

Evans, W. C. and Fuchs, G. (1988) Anaerobic degradation of aromatic compounds. *Annual Review of Microbiology* **42**, 289–317.

Gottschalk, G. and Andreesen, J. R. (1979) Energy metabolism in anaerobes. *Review of Biochemistry* **21**, 85–115.

Halliwell, B. (1984) Oxygen is dangerous: the nature and medical importance of oxygen radicals. *Medical Laboratory Sciences* **41**, 157–171.

Jones, W. J., Nagle, D. P. and Whitman, W. B. (1987) Methanogens and the diversity of archaebacteria. *Microbiological Reviews* **51**, 135–177.

Large, P. J. (1983) *Methylotrophy and Methanogenesis*. Wokingham: Van Nostrand Reinhold.

Mah, R. A., Ward, D. M., Baresi, L. and Glass, T. L. (1977) Biogenesis of methane. *Annual Review of Microbiology* **31**, 309–341.

Morris, J. G. (1975) The physiology of obligate anaerobiosis. *Advances in Microbial Physiology* **12**, 169–246.

Morris, J. G. (1976) Oxygen and the obligate anaerobe. *Journal of Applied Bacteriology* **40**, 229–244.

Morris, J. G. (1986) Anaerobiosis and energy-yielding metabolism. In: *Anaerobic Bacteria in Habitats Other than Man* (eds Barnes, E. M. and Mead, G. C.), pp. 1–21. Oxford: Blackwell Scientific Publications.

Odom, J. M. and Peck, H. D. (1984) Hydrogenase, electron-transfer proteins, and energy coupling in the sulfate-reducing bacteria *Desulfovibrio*. *Annual Review of Microbiology* **38**, 551–592.

Prins, R. A. (1977) Biochemical activities of gut micro-organisms. In: *Microbial Ecology of the Gut* (eds Clarke, R. T. J. and Bauchop, T.), pp. 73–183. London: Academic Press.

Rogers, P. (1986) Genetics and biochemistry of *Clostridium* relevant to development of fermentation processes. *Advances in Applied Microbiology* **31**, 1–60.

Rouvière, P. E. and Wolfe, R. S. (1988) Novel biochemistry of methanogens. *Journal of Biological Chemistry* **263**, 7913–7916.

Thauer, R. K., Jungerman, K. and Decker, K. (1977) Energy conservation in chemotrophic anaerobic bacteria. *Bacteriological Reviews* **41**, 100–180.

Thauer, R. K., Möller-Zinkhan, D. and Spormann, A. M. (1989) Biochemistry of acetate catabolism in anaerobic chemotrophic bacteria. *Annual Review of Microbiology* **43**, 43–67.

Thauer, R. K. and Morris, J. G. (1984) Metabolism of chemotrophic anaerobes: old views and some new concepts. In: *The Microbe 1984. Part II. Prokaryotes and Eukaryotes* (eds Kelly, D. P. and Carr, N. G.), pp. 123–168. Cambridge: Cambridge University Press.

Woods, D. R. and Jones, D. T. (1986) Physiological responses of *Bacteroides* and *Clostridium* strains to environmental stress factors. *Advances in Microbial Physiology* **23**, 1–64.

Zeikus, J. G. (1980) Chemical and fuel production by anaerobic bacteria. *Annual Review of Microbiology* **34**, 423–464.

6

Anaerobes of medical and veterinary significance

Diseases caused by obligate anaerobes have been recognized since antiquity. These ancient diseases are invariably of clostridial aetiology, since the clinical syndromes caused by the toxins of clostridia are sufficiently distinct to allow their identification from descriptions sometimes written many hundreds, if not thousands of years ago. For example, tetanus was described by Hippocrates almost 2500 years ago. This is not to say that disease caused by other anaerobes did not exist at the same time, but that the role of these other organisms in disease was not recognized until comparatively recently.

Many of the important clostridial infections, such as botulism and tetanus, are exogenous in origin. Gas gangrene follows infection of wounds with organisms which may be derived from the endogenous flora of the large intestine or from the environment. In contrast, the source of infection by non-sporing anaerobes is almost invariably endogenous. The infecting flora in these cases is derived from one of the habitats colonized by anaerobes, whether the oral cavity, the large intestine or the female genital tract.

Clostridial infections

The most important of the ancient clostridial diseases are botulism, tetanus and gas gangrene. These diseases result from the action of proteinaceous exotoxins that are amongst the most potent toxic materials known to man. The toxins of clostridia may be divided into three classes comprising neurotoxins, lethal and necrotizing toxins, and enterotoxins.

Several basic principles should be borne in mind when studying clostridial infections:

(i) Most, if not all toxigenic clostridia produce several toxic substances. The relative importance of each toxin often varies between strains of the same species. Not all strains of any one species will necessarily express all the toxins produced by that species, thus several species are subdivided into types according to their production of toxins.

(ii) Antigenic relationships between toxins produced by different species of clostridia are extremely common. Thus antitoxin against one species may partially or completely neutralize the activity of a toxin produced by a different species.

(iii) The nomenclature of clostridial toxins is complex. As the toxins of each species were identified, the general practice was to designate them with Greek characters. Thus *C. perfringens*, *C. novyi*, *C. septicum*, *C. histolyticum* and *C. sordellii* all produce α-toxins. Many of these α-toxins are the major toxins produced by each species, being lethal and necrotic. However, this is not always the case. The lethal and necrotizing toxin produced by pathogenic strains of *C. sordellii* is the β-toxin, but the α-toxin of *C. sordellii* is a lecithinase C which is antigenically similar to, but much less toxic than, the α-toxin of *C. perfringens*. The α-toxins of *C. septicum* and *C. histolyticum* are antigenically similar, yet the β-, γ- and δ-toxins of *C. septicum* and *C. chauvoei* are also very closely related. The clostridial toxins purified more recently have not been named in the same way. Thus the enterotoxin of *C. difficile* (toxin A) is similar to the haemorrhagic toxin (HT) of *C. sordellii*, while the lethal toxin of *C. sordellii* (LT) is antigenically and biologically related to cytotoxin (toxin B) of *C. difficile*.

(iv) The range of toxins produced by clostridia has yet to be fully determined, as have the biological and immunological relationships between them. With the application of modern biochemical techniques to the study of these compounds it is clear that there is much more to be learned about these molecules and their role in the pathogenesis of clostridial disease.

Botulism

Botulism is an intoxication caused by neurotoxins. Botulism almost invariably results from the ingestion of pre-formed toxin in food, although rarely botulism may follow wound infection by *C. botulinum*. Infant botulism follows infection of the infant gut by the organism and is discussed below.

Strains of *C. botulinum* produce one of several antigenically distinct neurotoxins, all of which exert very similar physiological effects when ingested. Six different botulinum toxin types are presently recognized. Botulinum neurotoxins are detected by mouse inoculation and are identified by the protective effects of type-specific antitoxin in mice.

Table 6.1 Toxin types of *Clostridium botulinum*

Group	Toxin type	Proteolytic	Fermentation of sucrose	End-products of glucose metabolism*
I	A B (proteolytic strains) F (proteolytic strains)	+	–	A,p,ib,B,iv,v,IC
II	E B (non-proteolytic strains) F (non-proteolytic strains)	–	+	A,B
III	C D	–	–	A,P,B

* A = acetic, p = propionic, ib = isobutyric, B = butyric, iv = isovaleric, v = valeric, IC = isocaproic acids. Capitals indicate major end-products and lower-case letters indicate minor end-products.

The foods recognized as being important vehicles of human botulism are generally those which are preserved with a minimum of heating and which are subsequently eaten either cold or with minimal re-heating. These include cured meats, bottled vegetables and canned products such as fish. The natural habitat of *C. botulinum* is soil and the organism may therefore contaminate vegetable products. Spores of *C. botulinum* survive heating relatively well whereas botulinum toxin is inactivated by heating at 80°C for 30 minutes or by boiling for 3 minutes.

Clostridium botulinum is not a biologically homogeneous species, but represents several biochemically distinct species characterized by their common production of botulinum toxin(s). The biochemical activities of the various toxin types of *C. botulinum* are summarized in Tables 3.4 and 6.1. Strains of toxin type A are proteolytic, as are some strains of type B and of type F. Strains of types C–E and the remaining strains of types B and F are saccharolytic but non-proteolytic. Food in which non-proteolytic *C. botulinum* has grown may not appear obviously spoiled for this reason.

Strains of *C. botulinum* type A and proteolytic strains of types B and F are biochemically identical to the non-pathogenic species *C. sporogenes*. There is some evidence that *C. sporogenes* strains can be converted to toxigenic strains of *C. botulinum* by infection with specific bacteriophage. Similarly, in strains of types C and D loss of toxigenicity has been associated with the elimination of a bacteriophage, re-infection with which restores the ability to produce toxin.

A further toxin type of *C. botulinum* (type G) was described in 1970. Botulism caused by this toxin type has not been reported and it was recently re-named *C. argentinense*. This organism is phenotypically similar to the non-toxigenic species *C. hastiforme* and *C. subterminale*. It remains to be seen whether re-naming of other toxin types of *C. botulinum* is justified on other than purely taxonomic grounds. However, three isolates of other *Clostridium* spp. have recently been recovered, from cases of infant botulism, that produce botulinum neurotoxins. Two isolates of *C. butyricum* were found to produce type-E toxin while an isolate of *C. barati* produced type-F botulinum toxin. The suggestion has been made that these normally non-pathogenic species may have acquired the ability to produce neurotoxins from strains of *C. botulinum*.

The neurotoxin(s) of *C. botulinum* are extremely potent: the lethal dose for man is estimated to be about 1 ng/kg body weight. The action of the toxin is to block the release of neurotransmitters from peripheral motor nerve endings. The clinical effect of this action is to induce flaccid paralysis. Botulinum toxin is adsorbed from the upper gastrointestinal tract and enters the blood, before becoming bound to cholinergic synapses in the peripheral nervous system. Once the toxin is bound to the site of action it remains fixed, thus the clinical symptoms of botulism cannot be reversed by the administration of antitoxin after their onset.

As with all food-borne intoxications the incubation period is short, generally less than 24 hours. Vomiting, thirst and transient paralysis of pharyngeal and ocular muscles occur, followed by flaccid paralysis of all voluntary muscles. Death may occur from respiratory failure. Recovery in non-fatal cases is protracted, the ocular muscles in particular remaining affected for many months.

The diagnosis of botulism is made on clinical grounds. Laboratory investigations are made in order to confirm the diagnosis and to trace the source of the intoxication. Although antitoxin has no effect on toxin already bound to nerve endings it can prevent binding of further toxin and thus it is administered as soon as possible after a clinical diagnosis has been made. Other measures, such as respiratory support, are taken to provide symptomatic relief. Antitoxin may be administered prophylactically to individuals who may have been exposed to botulinum toxin at the same time as the patient (by sharing a meal). Laboratory investigations include the detection of toxin in suspect food, faeces and serum using the mouse assay, and the isolation of the organism (and demonstration of its toxigenicity) from food and faeces.

Prevention of botulism is achieved by rigorous quality control of food-processing plants and by the inclusion in foodstuffs of inhibitory agents such as nitrate and sodium chloride, and the maintenance of a low pH, in order to prevent the germination of spores of *C. botulinum* which may remain in processed food. The risk of botulism is much greater when foods are preserved at home by methods other than freezing. Persons at increased risk of exposure to botulinum toxin, such as laboratory workers, are normally immunized with a polyvalent toxoid vaccine.

Wound botulism follows growth of *C. botulinum* in infected wounds with concomitant production of neurotoxin which is absorbed directly from the wound. This type of botulism is extremely rare; all cases so far have followed infection with *C. botulinum* of type A or type B.

Botulism also affects animals, but different species vary widely in their susceptibility to the neurotoxin. Among animals of economic importance to man, cattle and sheep are affected by naturally occurring botulism. Vast outbreaks of botulism affect wildfowl in late summer of some years, as the water level in lakes falls and exposes mud containing spores of *C. botulinum* type C. Growth and toxin production occurs in dead crustaceans, which are then eaten by birds of many species. Typically, thousands of birds die, and their putrefying carcasses release more spores of *C. botulinum* into the mud. Pheasants and other game birds also suffer from botulism after eating maggots containing botulinum toxin. It is of ecological interest that maggots and vultures appear to be quite resistant to the effects of botulinum neurotoxin.

Tetanus

Tetanus is also caused by a neurotoxin of high potency, with a lethal dose for man of about 2.5 ng/kg. The neurotoxin of *C. tetani* is known as tetanospasmin and its clinical effect is to produce tetany or rigid spasms of voluntary muscles. Unlike botulism tetanus always is the result of infection of wounds by *C. tetani*. In contrast to the clostridia that cause gas gangrene, *C. tetani* is neither histotoxic nor invasive and therefore a wound infected with *C. tetani* may show little or no sign of infection.

The habitat of *C. tetani* is soil and it is ubiquitous in distribution. The organism is also isolated not infrequently from the faeces of animals and man. Before the aetiology of tetanus was understood the disease typically followed either agricultural or gardening accidents or wounds sustained in battle. A more common presentation of tetanus nowadays is in an elderly person with a history of a minor wound, such as a prick from a rose thorn sustained while gardening, which apparently healed uneventfully without the need for medical treatment. Such patients often have a record of inadequate immunization. Tetanus in urban areas may also follow unhygienic injection of drugs of abuse.

In some developing countries however, tetanus remains a major public health problem, and it has been estimated that 500 000 deaths result from tetanus annually, most of whom are newborn infants suffering neonatal tetanus. Neonatal tetanus is almost always contracted following unhygienic cutting of the umbilical cord and staunching of the blood flow with dung or other such material, and has a mortality rate exceeding 90%. Puerperal and post-abortal tetanus also occur in significant numbers, again being related to unhygienic practices at the time of birth or abortion.

Not all wounds contaminated with *C. tetani* will lead to the development of tetanus in non-immune individuals. *Clostridium tetani* is an exacting anaerobe and spores will not germinate unless conditions within the wound are favourable for growth of the organism. The principal requirement for growth is a low redox potential. This is not found in healthy, well-oxygenated tissues, which usually have an Eh greater than 100 mV, but may be brought about by the presence in wounds of necrotic tissues, blood clots, foreign bodies and other bacteria. The interruption of blood flow to a wounded area will also contribute to a fall in Eh. These conditions will often be found in major wounds but rarely in minor wounds. Moreover, even when conditions favour growth, a high proportion of strains of *C. tetani* are non-toxigenic.

The incubation period of tetanus varies widely, depending upon the severity of the wound and the size of the inoculum of *C. tetani*, but is usually about a week. Tetanospasmin is produced by *C. tetani* as the organism grows in the wound. The toxin inhibits the release of inhibitory transmitters from synapses. It shares this presynaptic action with botulinum neurotoxin. The rigid paralysis (tetany) characteristic of tetanus results from the action of tetanospasmin upon the central nervous system. The toxin is thought to reach the central nervous system by axonal transport up motor nerves. Tetanospasmin may also affect neuromuscular transmission in a manner similar to that of botulinum neurotoxin, but the relative specificity of tetanospasmin for inhibitory synapses in the central nervous system is much greater.

Tetanus generally begins with the development of trismus (lockjaw) and dysphagia. Within a few hours these symptoms progress to a characteristic grinning expression (risus sardonicus). As the course of the disease proceeds, reflex spasms involving all voluntary muscles occur. In severe cases they first appear very soon after the onset of symptoms, but in mild cases they may not appear for several days. Spasms may be induced by external stimuli such as movement of the patient and loud sounds, but later in the disease they occur spontaneously. As the muscles of the back are the most powerful in the body the patient will usually assume an arched position due to hyperextension of the spine. The jaws are tightly clenched and the head bent backwards. The limbs are usually extended rigidly. This characteristic position is known as opisthotonos. During the spasms, which may last for several minutes in severe cases, breathing stops. Throughout this unpleasant experience the patient remains fully conscious. Death may occur due to respiratory or cardiac failure.

The diagnosis of tetanus is made solely on clinical grounds. If an obvious wound is present then laboratory confirmation may be possible by isolation of *C. tetani* and demonstration of toxigenicity in mice. Isolation of toxigenic *C. tetani* from a wound in the absence of symptoms does not mean that a patient has tetanus, nor that they will necessarily develop the disease.

As is also the case in botulism, tetanospasmin cannot be neutralized by antitoxin once it becomes fixed to nerve cells. However, antitoxin administered at this stage will prevent further binding of tetanospasmin and thereby will help to reduce the severity of tetanus. Other therapeutic measures are aimed at controlling reflex spasms, maintaining adequate nutrition and ventilation (via a tracheostomy) and the prevention of aspiration pneumonia. Surgical cleansing of obvious wounds is of paramount importance in order to remove the source of toxin and prevent further growth of *C. tetani*.

Prevention of tetanus is accomplished by immunization with tetanus toxoid. It is important that regular booster injections should be given at 5–10-year intervals in order to maintain adequate immunity, since exposure may occur at any time of life. Patients with wounds should have surgical debridement, and passive immunization with antitoxin is necessary for non-immune individuals.

Gas gangrene

In contrast to tetanus and botulism, which are monospecific infections, gas gangrene follows wound infections which are almost invariably polymicrobial. The organisms responsible are a group of histotoxic clostridia that produce a range of lethal and necrotizing toxins, in addition to a variety of other virulence factors. These organisms, often referred to as the gas gangrene bacilli, are listed in Table 6.2.

Because of the nature of the contamination of wounds that may lead to the development of gas gangrene, it is usual to find several other species of anaerobes and facultative anaerobes in such wounds. The anaerobes present will often include non-sporing anaerobes and non-pathogenic clostridia, such as *C. bifermentans*, *C. fallax*, *C. sporogenes* and *C. tertium*. *Clostridium perfringens* type A is by far the most common cause of gas gangrene, being recovered from up to 80% of cases. *Clostridium novyi* type A may be isolated from more than a third of cases, while *C. septicum* is associated with a smaller proportion of cases. Other histotoxic clostridia are of lesser importance in gas gangrene in man; *C. chauvoei* is exclusively an animal pathogen.

As noted in the section on tetanus, not all wounds contaminated with histotoxic clostridia will necessarily give rise to the development of gas gangrene. *Clostridium perfringens* is particularly important in this respect, since it is one of the most aerotolerant clostridia and one of the most rapidly

Table 6.2 Histotoxic clostridia

Species	Disease	Species affected
C. chauvoei	blackleg	cattle, sheep
C. histolyticum	gas gangrene	man
C. novyi type A	gas gangrene	man
type B	necrotic hepatitis (black disease)	sheep
type D	bacillary haemoglobinuria (redwater disease)	cattle
C. perfringens type A	gas gangrene	man
type C	struck	sheep
C. septicum	gas gangrene	man
	blackleg	pigs
	malignant oedema	cattle
C. sordellii	gas gangrene	man

growing. The maximum Eh at which *C. perfringens* will grow in healthy muscle at pH 7.5 is approximately 40 mV; as we have noted above, the Eh in healthy tissues is always greater than 100 mV. However, if the pH falls to 6.5 or below, as happens in muscle which has been anoxic for 4 hours or more, then *C. perfringens* will grow at an Eh of about 160 mV. Clearly, once these conditions obtain and growth of *C. perfringens* commences, the Eh and pH will rapidly fall and other clostridia will also be able to grow.

Three stages of clostridial wound infection are recognized, of increasing severity. The first stage is one of simple contamination, where one or more of the gas gangrene bacilli may be isolated from the wound but there is no evidence of infection of surrounding tissues. At this stage, treatment consists of surgical debridement and administration of penicillin, but if the wound is left untreated then conditions may favour the development of the second stage, clostridial cellulitis. Here there is invasion of the fasciae near the wound but not of muscle tissue itself. At this stage there is little or no toxin production, since the flora is usually composed of weakly toxigenic organisms. The infection develops quite slowly, there is little or no oedema, but a seropurulent discharge is produced and gas is present in the wound. The final condition is that of clostridial myonecrosis (gas gangrene), in which healthy muscle tissue is invaded, accompanied by the production of clostridial exotoxins.

The incubation period for gas gangrene may be as little as 6 hours, depending upon the severity of the wound and the degree of contamination. Initially, there is pain in the wounded region, accompanied by increasing oedema. Both the pulse rate and temperature rise. A copious serosanguinous discharge exudes from the wound. The overlying skin becomes taut and discoloured and is covered with vesicles. As the infection progresses bubbles of gas escape from the wound and crepitation is felt within the tissues. The infection is fulminating and the patient rapidly goes into shock, followed by delirium, coma and death.

Diagnosis of gas gangrene is made on clinical grounds and emergency treatment necessitates amputation of the affected limb or other region when possible. Once gas gangrene has become established the prognosis is extremely poor. The mortality has been reduced somewhat in recent years by the use of hyperbaric oxygen therapy when available. However, the condition is entirely preventable, provided adequate surgical treatment of wounds is readily available. During the 1914–18 war it is estimated that 100 000 German soldiers died from gas gangrene; during the eight years of American involvement in Vietnam, less than 30 cases were recorded.

Disease in animals caused by the histotoxic clostridia may take several clinical forms, depending upon the species and toxin type responsible. Animal infections also tend to be endogenous, the organisms gaining entry to healthy tissues following translocation from the gut. The most common presentation in cattle is blackleg, caused by *C. chauvoei*. This condition affects cattle which are well fed and in excellent health; malnourished or heavily parasitized animals appear resistant to blackleg. The onset of blackleg resembles that of gas gangrene and the disease progresses to a fatal outcome within 12–36 hours. The principal lesion almost always occurs in large muscle blocks such as the shoulder, thigh or neck.

Another syndrome affecting cattle specifically is bacillary haemoglobinu- ria, caused by *C. novyi* type D (*C. haemolyticum*). In this condition the primary focus of infection is the liver. Destruction of liver tissue by immature liver flukes (*Fasciola hepatica*) burrowing in search of bile ducts provides conditions favourable for the growth of *C. novyi* type D. Release of *C. novyi* type D β-toxin into the bloodstream causes lysis of circulating erythrocytes and the destruction of the capillary epithelium. The large amounts of haemoglobin released are excreted in the urine, hence the common name for this condition (redwater disease). Death of the animal occurs about three days after the onset of infection. Burrowing by *F. hepatica* also predisposes to *C. novyi* type-B infection in sheep, known as necrotic hepatitis. Growth of *C. novyi* type-B in the necrotic regions of the liver is accompanied by the production of the lethal α-toxin of *C. novyi*. The descriptive name of black disease refers to the appearance of the liver at autopsy.

In addition to these endogenous infections there is a number of clostridial wound infections which affect domestic animals. Malignant oedema in cattle is analogous to gas gangrene caused by *C. septicum* in man, while *C. chauvoei* is a cause of blackleg in sheep following wounds. Shearing wounds, castration, tail docking and even umbilical stumps are all recognized as potential sites of entry of *C. chauvoei*.

Diagnosis of these conditions is often made at autopsy; this in itself may be difficult because of the rapidity with which clostridia such as *C. septicum* invade the tissues after death. Moreover, spores of *C. chauvoei* are often found in healthy livers. Thus, the isolation of histotoxic clostridia from necrotic animal tissue is not necessarily diagnostic. Immunofluorescent microscopy is more valuable since it allows the detection and identification of numerically dominant organisms in fresh samples of necrotic tissue. Infections in farm animals caused by *C. chauvoei*, *C. novyi* and *C. septicum* are of sufficient economic importance to justify their control by widespread immunization using polyvalent toxoids.

Clostridial enteropathies

A large number of clinically distinct enteropathic syndromes is caused by several *Clostridium* spp. (Table 6.3). Some of these conditions were first described in the early years of bacteriology, but several of them have been recognized and associated with a clostridial aetiology only within the last 15 years. The majority of gut diseases caused by clostridia occur only in association with some predisposing condition. This may be:

(i) an immature gut flora which cannot prevent colonization (infant botulism, neonatal diarrhoea in hares, lamb dysentery, necrotic enteritis in piglets, iota-enterotoxaemia in rabbits);

(ii) a disturbance of the normal gut flora by antibiotics, removing the normal colonization resistance of the gut (antibiotic-associated diarrhoea and colitis);

(iii) an abrupt change in diet, usually associated with gluttony, which induces stasis of the bowel (pigbel, braxy);

(iv) an impairment of the immune system (neutropenic enterocolitis).

Infant botulism occurs when the large bowel of the infant is colonized by *C. botulinum*, followed by production of botulinum neurotoxin *in vivo*. The symptoms of infant botulism are usually not as severe as those of food-borne botulism, but the organism has been implicated as one cause of the sudden infant death syndrome (SIDS). Infant botulism was first described in the USA but has since been reported in several other countries. It is now recognized as the most common form of botulism. Attempts to discover the

Table 6.3 Clostridial enteropathies

Species	Disease	Species affected
C. botulinum type A	infant botulism	man
type B	infant botulism	man
	shaker foal syndrome	horses
type C	botulism	chickens
C. colinum	ulcerative enteritis (quail disease)	quails, pheasants and chickens
C. difficile	antibiotic-associated diarrhoea and pseudomembranous colitis	man
	neonatal diarrhoea	hares
C. perfringens type A	food poisoning, antibiotic-associated diarrhoea	man
type B	lamb dysentery (necrotic enteritis)	sheep
type C	pigbel (necrotic enteritis)	man
	necrotic enteritis	piglets
type D	enterotoxaemia	sheep
C. septicum	enterotoxaemia (braxy)	sheep
	neutropenic enterocolitis	man
C. spiroforme	iota-enterotoxaemia	rabbits

source of *C. botulinum* in this condition led to the finding of *C. botulinum* spores in many brands of honey in the USA. Careful case-control studies have failed to confirm that honey is a vehicle of infection in infant botulism. Most cases of infant botulism have occurred in the western states of the USA, where *C. botulinum* types A and B are commonly distributed in soil. Shaker foal syndrome is a condition occurring in young horses, analogous to infant botulism in humans.

The role of *C. difficile* in antibiotic-associated diarrhoea (AAD) and pseudomembranous colitis (PMC) was first recognized in 1977, since when the organism and its toxins have been the subjects of much intensive study. *Clostridium difficile* is an uncommon component of the normal adult gut flora; its normal habitat appears to be the gut of human infants and the young of other animal species. Infection of the adult gut may occur if the protective effect of the normal flora is compromised, usually by antibiotic therapy. The organism produces an enterotoxin (toxin A) and a cytotoxin (toxin B) which appear to act together to elicit a spectrum of symptoms which range from mild, self-limiting diarrhoea to profuse watery diarrhoea accompanied by the presence of a fibrinous pseudomembrane overlying the mucosal surface of the large intestine. By its very nature this disease is often a hospital-acquired (nosocomial) infection and outbreaks of AAD may occur in hospital wards. It is ironic that this condition responds to treatment with antibiotics.

It is still not understood why this organism inhabits the infant gut and produces large amounts of both toxin A and toxin B without the infant usually suffering any pathological effects. One possibility is that the infant gut mucosal surface lacks the receptors for either or both toxins.

Considering the variety of toxins produced by *C. perfringens* and the ubiquity of its presence in the gut of man and animals, it is hardly surprising that a wide variety of enteric diseases is caused by this organism. *Clostridium perfringens* type A is one of the most common causes of food poisoning. The symptoms of this condition result from the action of an enterotoxin liberated in the small intestine as vegetative cells of some *C. perfringens* strains sporulate. A mild diarrhoeal disease usually follows a few hours after ingestion of the contaminated food. A prerequisite for this condition is the presence in food of large numbers of cells of *C. perfringens*, since the yield of enterotoxin is proportional to the number of sporulating cells. *Clostridium perfringens* food poisoning therefore always follows consumption of food which has been allowed to rest at room temperature (permitting multiplication of *C. perfringens*) and which subsequently has been re-heated inadequately before serving. Meat dishes, including large joints and stews, are often the vehicle of infection; food poisoning caused by *C. perfringens* often occurs in institutions such as prisons, hospitals and schools, and among guests at receptions or customers at restaurants. A knowledge of the aetiology of this condition leads to the conclusion that it is preventable by hygienic practices within the kitchen. Enterotoxigenic *C. perfringens* type A also has been shown to cause a few cases of AAD in elderly hospital patients.

Of the remaining enteropathies caused by *C. perfringens*, gluttony is a common precipitating factor (it might be said that gluttony also plays a role in *C. perfringens* food poisoning). Only the greediest lambs are affected by lamb dysentery, which occurs during the first two weeks of life. In this condition, *C. perfringens* type B invades the small intestinal wall and causes extensive necrosis and gas formation within the submucosal layer (effectively gas gangrene of the bowel), often leading to perforation and peritonitis. Enterotoxaemia caused by *C. perfringens* type D often follows a change from poor to rich pasture.

Gluttony is similarly an underlying factor in necrotic enteritis (pigbel) affecting man. This condition is largely, but not entirely, restricted to hill tribes in Papua-New Guinea. The members of this society subsist on a diet rich in sweet potatoes, which contain potent inhibitors of trypsin. At times of communal celebration large numbers of pigs are slaughtered and cooked on coals in shallow pits. *Clostridium perfringens* type C is commonly found in the gut of pigs and the communal method of slaughtering ensures that the organisms are liberally distributed on the meat, which is then cooked in large joints. When the meat is cooked the participants at the feast gorge themselves on pork.

It is perhaps not surprising that a combination of these factors (contaminated meat, inadequate cooking, gluttony and lack of dietary enzymes) leads to intestinal stasis, followed by multiplication of *C. perfringens* and invasion of the wall of the small intestine. The course of the disease is usually fatal unless the affected portions of the bowel are removed surgically.

Clostridium septicum is the cause of an enterotoxaemia in sheep known as braxy. This condition affects lambs in their first season, in areas which have heavy frosts. Ingestion of frozen grass chills the wall of the abomasum and allows the organism to invade through areas of localized frostbite. Death follows the development of septicaemia and profound toxaemia. In man, an association between *C. septicum* infection and gastrointestinal malignancies has been recognized for many years. More recently, invasive *C. septicum* enterocolitis in neutropenic patients has been osbserved. Treatment consisting of surgical resection of affected portions of the bowel is sometimes effective in this otherwise fatal condition.

Weanling diarrhoea in rabbits was originally ascribed to infection with *C. perfringens* type E, because of the detection of a toxin neutralized by iota-antitoxin in the faeces of affected animals (iota toxin production characterizes *C. perfringens* type E). However it was found that *C. spiroforme* was responsible for this condition. This represents another example of cross-reactivity between toxins produced by two *Clostridium* spp.

Because of the rapidly progressive nature of many of these infections, particularly those invasive enteropathies caused by *C. perfringens* and *C. septicum*, treatment is confined to surgical removal of the infected parts of the gastrointestinal tract. Control of these infections is generally achieved by immunization with toxoid vaccines. This is a successful strategy against the conditions affecting domestic animals and also against pigbel in man.

Infections caused by non-sporing anaerobes

The occurrence of non-sporing anaerobic infections reflects the distribution of these organisms in the normal flora of the oropharynx, the large intestine and the female genital tract. The predominance of *Bacteroides* spp. in the human intestinal flora was noted over 50 years ago, yet the important role played by these organisms in many types of infection was not clearly

Table 6.4 Some non-sporing anaerobic infections of man

Head, neck and upper respiratory tract
Brain abscess
Dental abscess
Periodontal disease
Vincent's disease (necrotizing gingivitis)
Tonsillitis
Otitis media
Lung abscess
Aspiration pneumonia

Gastrointestinal tract
Abdominal wound infections
Perianal abscess
Liver abscess
Subphrenic abscess
Peritonitis
Synergistic gangrene (necrotizing fasciitis)

Female genital tract
Bacterial vaginosis
Endometritis
Septic abortion
Pelvic inflammatory disease
Pyometra
Bartholin's abscess
Wound infections
Peritonitis

Other sites
Balanoposthitis
Breast abscess
Human bite infections
Infection of chronic ulcers
Infection of tumours
Necrobacillosis
Paronychia
Penetrating wound infections
Septicaemia
Tropical ulcer

---

understood until comparatively recently. The failure to recognize the potentially pathogenic nature of non-sporing anaerobes was due, at least in part, to the technical difficulties associated with anaerobic work. However,

Table 6.5 Predisposing factors in endogenous anaerobic infections

---

Cardiovascular disease
Childbirth and abortion
Chronic and malignant disease
Decubitus and varicose ulcers
Diabetes mellitus
Genital tract infection
Human bites
Intrauterine contraceptive device use
Surgery and trauma

---

the failure of these organisms to elaborate exotoxins and therefore the lack of distinct clinical syndromes associated with their production must also have contributed to the lack of interest shown in non-sporing anaerobic infections. Moreover, it is almost invariably the case that non-sporing anaerobic infections are polymicrobial and involve both facultative and obligate anaerobes. Thus it was accepted for many years that these infections were primarily caused by facultative anaerobes, particularly Gram-negative, fermentative bacilli such as *Escherichia coli*.

The wide variety of human infections which may be due to non-sporing anaerobes is summarized in Table 6.4. Similar infections occur in animals. Although the species involved may differ the source of the infecting organisms and the predisposing factors are often identical to those observed in human infections. For these reasons, human and veterinary infections will be considered together.

Since these infections are almost overwhelmingly endogenous in origin they may be subdivided according to the source of the infecting flora. Thus infections of the head, neck and upper respiratory tract are primarily derived from the oropharyngeal flora, those of the internal organs are usually of intestinal origin and infections of the female genital tract are derived from the normal flora of the vagina. Anaerobic surgical wound infections may follow surgery on any of the sites with an anaerobic flora. Similarly septicaemia may originate from a primary focus of infection associated with any of these sites; the isolation of non-sporing anaerobes from blood cultures may indeed be a guide to the source of infection in patients with an unexplained fever.

Not all non-sporing anaerobic infections are derived from the patient's own normal flora, but the exceptions are endogenous in that the infecting organisms originate from the normal flora of another individual. Breast abscesses in non-lactating women frequently follow human bites on the nipples. Other anaerobic infections of human bite wounds include those following injury to the knuckles during fist-fights. Anaerobic balanoposthitis is a purulent infection of the foreskin and glans penis, which occurs in males with poor hygienic habits; the infecting organisms are derived from the vaginal, oral or intestinal flora of sexual partners.

The factors which predispose to endogenous anaerobic infection are many. Some important ones are summarized in Table 6.5. Some illustrative non-sporing anaerobic infections are synergistic. These include Vincent's disease, bacterial vaginosis and ovine footrot.

Vincent's disease

This condition is also known as acute ulcerative gingivitis, necrotizing gingivitis and trench mouth. It generally occurs in patients with poor oral hygiene; similar conditions occur in a number of animal species. The symptoms of Vincent's disease include halitosis, painful and bleeding gums, and pseudomembranous ulceration of the gums, leading to destruction of tissues supporting the teeth, including the gums, the periodontal ligaments and alveolar bone.

Vincent's disease was traditionally regarded as a synergistic fuso-spirochaetal infection, involving *Fusobacterium nucleatum* and a spirochaete, *Borrelia vincentii*. The involvement of other oral anaerobes such as anaerobic cocci, black-pigmented *Bacteroides* spp. and *Actinomyces odontolyticus* is also probable. Vincent's disease may be diagnosed on clinical grounds, supported by the observation of a characteristic mixed bacterial flora in Gram-stained smears of the lesions. The condition responds rapidly to therapy with metronidazole, but bone and gum loss are irreversible.

Bacterial vaginosis

Bacterial vaginosis is a common, non-inflammatory infection of the vagina. It is characterized by the presence of a thin, homogeneous, greyish-white discharge. Laboratory investigation reveals a vaginal pH ≥ 5, the presence of clue cells (vaginal epithelial cells densely coated with bacteria), a ratio of succinate:lactate raised to 0.4 or greater, a reduced vaginal redox potential (Eh) and the absence of pus cells; amines, including putrescine and cadaverine, are also present. These findings are summarized in Table 6.6.

Bacterial vaginosis was formerly termed non-specific vaginitis and was considered to result from infection of the vagina by *Gardnerella vaginalis*. A more recent synonym is anaerobic vaginosis. Bacterial vaginosis is ubiquitous; in some populations bacterial vaginosis may represent a quarter of cases presenting at genitourinary clinics and one third of all diagnoses made

Table 6.6 Summary of clinical and laboratory findings in bacterial vaginosis

Parameter	Normal vagina	Bacterial vaginosis
Discharge	absent	present
Vaginal pH	<4	>5
Eh	150–200 mV	−90 mV
Pus cells	absent	present (few)
Clue cells	absent	abundant
Volatile acids (succinate : lactate)	<0.4	>0.4
Amines	absent	present
Anaerobes	present	numerous
Gardnerella vaginalis	absent*	numerous

* Asymptomatic carriage may occur.

therein. A diagnosis of bacterial vaginosis may be made by the detection of three of the following: characteristic discharge, pH $\geqslant$ 5, clue cells and a positive test for amines using potassium hydroxide. Culture of vaginal discharge is not necessary to effect a diagnosis. Bacterial vaginosis responds readily to treatment with metronidazole.

Bacterial vaginosis results from the synergistic interaction of *G. vaginalis* and obligate anaerobes, including *Bacteroides* spp. commonly found in low numbers in the normal vaginal flora, such as *B. bivius*, *B. disiens* and *B. asaccharolyticus*, anaerobic Gram-positive cocci and Gram-negative curved rods (*Mobiluncus* spp.). The interactions between *G. vaginalis* and the obligate anaerobes involved in bacterial vaginosis are shown in Fig. 6.1.

It is probable that the large amounts of succinate produced contribute to the symptoms of bacterial vaginosis and the absence of an inflammatory response. Succinic acid is the major metabolic end-product of *Bacteroides* spp. and *Mobiluncus* spp. and has been shown to inhibit polymorph migration *in vitro* at pH 5.5. The *Bacteroides* spp. isolated from patients with bacterial vaginosis are powerful producers of amines, principally cadaverine and putrescine, which give rise to the positive KOH amine test. In addition, amine production also results in a rise in vaginal pH.

Optimum growth of *Mobiluncus* spp. occurs when the pH is > 5, while the optimum pH for growth of *G. vaginalis* is 6–6.5. Thus, the production of amines by *Bacteroides* spp. stimulates the growth both of *G. vaginalis* and of *Mobiluncus* spp. *Gardnerella vaginalis* releases both pyruvate and amino acids during growth *in vitro*, which are metabolized by the anaerobic vaginal

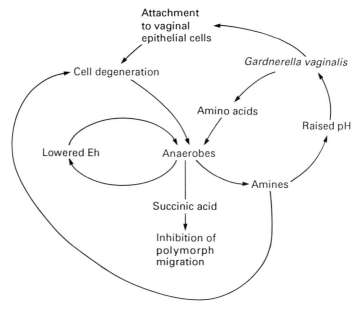

Fig. 6.1 Synergism in bacterial vaginosis.

flora with the production of amines. The inhibition of amine production by metronidazole may account for the apparently favourable effect of metronidazole on *G. vaginalis* (which is not an obligate anaerobe), since when amine production is halted by metronidazole, vaginal pH will fall to normal levels; growth of *G. vaginalis* is inhibited at ⩽ pH 4.5. Maximum adhesion of *G. vaginalis* to vaginal epithelial cells occurs between pH 5 and pH 6. Thus, the fall in vaginal pH associated with the cessation of amine production accounts also for the disappearance of clue cells consequent upon successful treatment. It is clear that bacterial vaginosis cannot be considered a communicable disease, but should be regarded as a perturbation of the ecology of the vaginal microflora.

Ovine footrot

Footrot is a disease of cloven-hoofed animals that is of considerable economic importance. Infection of the epithelium underlying the hoof results in an inflammatory response, which leads to the separation of the hoof. Lameness and inability to stand result in the animal struggling to feed by resting on its knees. Footrot spreads in moist, warm conditions. Ovine footrot appears to be a synergistic infection, primarily involving *Bacteroides nodosus* and *Fusobacterium necrophorum*. Under suitable conditions, *F. necrophorum* may cause superficial lesions in the interdigital space, which are then invaded by *B. nodosus*, followed by penetration of the keratinized tissue of the hoof. This is in turn followed by further invasion by *F. necrophorum*. Sloughing of the horn from its underlying tissue results from the inflammatory response induced by *F. necrophorum*. A vaccine containing pili of *B. nodosus* is protective, but the involvement of other, as yet unidentified anaerobes, in the pathogenesis of this condition is almost certain. Aerobes such as *Corynebacterium pyogenes* may also play a role, in providing conditions favourable to the initial colonization by *F. necrophorum*.

Footrot in many other animals also involves *F. necrophorum*. Two types of footrot occur in cattle, one analogous to ovine footrot, while the other appears to be a synergistic infection by *F. necrophorum* and *C. pyogenes*.

Necrobacillosis

Although many non-sporing anaerobic infections are polymicrobial, a few are monospecific in aetiology. The most important of these conditions in man is necrobacillosis, an acute infection with *F. necrophorum*, which usually follows a sore throat in otherwise healthy individuals. The symptoms of severe sore throat are followed by the sudden onset of fever and rigors. Metastatic abscesses develop, following haematogeneous spread of the organism, producing symptoms of pulmonary infection, osteomyelitis and purulent arthritis. Recovery follows appropriate antibiotic therapy (usually penicillin and/or metronidazole). Necrobacillosis also occurs in animals, the clinical symptoms including liver abscess in adult cattle, labial necrosis in rabbits and calf diphtheria.

Swine dysentery

Swine dysentery affects weaned piglets, particularly in intensive fattening units, and is characterized by bloody diarrhoea. It is caused by a spirochaete, *Treponema hyodysenteriae*. The symptoms result from invasion of the mucosal surface of the large bowel by the organism. Protection against swine dysentery may be achieved by immunizing piglets with inactivated *T. hyodysenteriae*.

Virulence factors of obligate anaerobes

Obligate anaerobes produce a considerable number of virulence factors, in addition to a wide spectrum of less overt substances which nevertheless contribute to their virulence. Among the obvious virulence factors are the lethal and necrotizing toxins and the enterotoxins of the pathogenic clostridia. The role of some other extracellular products of clostridia in pathogenicity is less clear, but it is likely that they contribute to virulence to a lesser extent. These compounds include haemolysins and enzymes such as neuraminidases, histidine decarboxylase, deoxyribonucleases, fibrinolysin, hyaluronidases, collagenases, gelatinases, lipases and elastases. Many of these enzymes may be important in facilitating the spread of histotoxic clostridia through healthy tissues. Of the clostridia, only *C. perfringens* produces a capsule.

The virulence factors of non-sporing anaerobes have been studied less closely than those of clostridia. All Gram-negative anaerobes produce endotoxin, the effects of which are very similar to those of endotoxin from other Gram-negative bacteria. The endotoxin of *F. necrophorum* appears to be particularly active, having several times the activity of *B. fragilis* endotoxin, for example. *Fusobacterium necrophorum* produces fibrinolysin, a haemolysin, deoxyribonuclease, gelatinase, lipase, caseinase, a leukocidin and a cytoplasmic toxin in addition to endotoxin.

By far the most important virulence factor of *B. fragilis* is its polysaccharide capsule. The capsular material itself induces abscess formation, whether in conjunction with cells of *B. fragilis* or in a purified form. A second effect of the capsule is the inhibition of phagocytosis. *Bacteroides fragilis* strains which do not produce capsules do not induce abscesses and are relatively avirulent. *Bacteroides fragilis* also produces a haemolysin, fibrinolysin and a range of enzymes including neuraminidase, deoxyribonuclease, phosphatase, hyaluronidase, heparinase, chondroitin sulphatase, collagenase and several other proteases. Some strains of *B. fragilis* elaborate an enterotoxin which causes diarrhoea in calves, piglets, foals and man.

Proteolytic enzymes produced by *Bacteroides* and *Fusobacterium* species may also contribute to pathogenicity in mixed infections. For example, collagenase produced by *B. melaninogenicus* enhances the effect of *F. necrophorum* infection in rabbits. Similarly, many non-sporing anaerobes including *B. fragilis*, *B. melaninogenicus*, fusobacteria and anaerobic cocci produce elastase. This enzyme may play a role in the involvement of non-sporing anaerobes in lung infections such as chronic bronchitis.

β-lactamase can also be considered a virulence factor. Among the anaerobic organisms that produce β-lactamase are *B. fragilis*, *B. bivius*, *B. disiens*, *B. oralis*, *F. nucleatum*, *C. butyricum*, *C. ramosum* and several of the black-pigmented *Bacteroides* spp. The involvement of *B. melaninogenicus* and similar *Bacteroides* spp. in the upper respiratory tract flora may lead to failure of penicillin therapy of streptococcal sore throat.

Metabolites of anaerobes also contribute to virulence, although the extent to which this occurs *in vivo* is unknown. Succinic acid inhibits polymorph migration *in vitro* at pH 5.5. This may partly explain the absence of an inflammatory response in bacterial vaginosis. It has been suggested that the amines produced in bacterial vaginosis also play a direct role in pathogenesis. Similarly butyric acid is responsible for a cytotoxic effect on tissue cultures exhibited by *C. butyricum*.

The role of adhesion in the pathogenesis of anaerobic infection has not been studied extensively. Adhesion of enteropathogenic clostridia to intestinal mucosae has been demonstrated *in vivo*. *Fusobacterium nucleatum* and black-pigmented *Bacteroides*, organisms involved in periodontal disease, adhere to gingival cells. One problem facing those who would study adhesion of anaerobes *in vitro* is the maintenance of viable models in an anaerobic environment, or vice versa the maintenance of viable anaerobes in an atmosphere favourable to living animal cells.

Diagnosis of anaerobic infection

Infections caused by obligate anaerobes of all kinds are characterized by the presence of a fetid odour, resulting from the production of volatile metabolites by the anaerobes within the lesion. The compounds chiefly responsible are short-chain volatile acids in the series acetic–caproic, but other metabolites such as amines, and volatile aromatic compounds such as indole and skatole undoubtedly contribute to the overall effect. Thus, a lesion with a distinctive malodour can be assumed to be anaerobic in aetiology.

Volatile fatty acids may be detected in pus and other liquid samples by smell or by gas–liquid chromatography (GLC; see Chapter 2). This method of detecting anaerobic infection has the advantage that the metabolic products remain in the specimen even if the organisms have perished during transit to the laboratory. However, it is not possible to predict the identity of the infecting anaerobes by direct GLC of clinical material. Examples of GLC traces from anaerobic pus are shown in Fig. 6.2. Gas chromatography may also be used to detect anaerobic growth in blood cultures, but is of no value in detection of anaerobic infections of the gut (such as the clostridial enteropathies), since faeces contain the metabolic products of the normal bowel flora.

Another rapid test that may be helpful in the laboratory diagnosis of anaerobic infection is the detection of red fluorescence under long-wave ultraviolet light at a wavelength of 365 nm (Wood's lamp). This is characteristic of the black-pigmented *Bacteroides* spp. and the detection of red fluorescence in pus, on dressings or in other specimens indicates infection by one of these species, and by inference, with other anaerobes also. A variety of other rapid diagnostic tests for obligate anaerobes has been developed. Specific immunofluorescent stains facilitate the detection of *B. fragilis*,

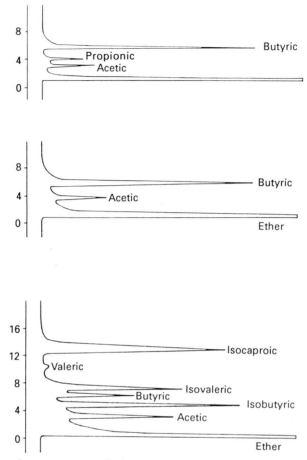

Fig. 6.2 Gas chromatograms of ether extracts of pus from anaerobic infections, containing volatile acid products of anaerobic metabolism (see also Fig. 2.5).

B. melaninogenicus, *C. septicum* and *C. chauvoei* in clinical material. Detection of clostridial toxins in faeces may be accomplished by neutralization of cytotoxic effects by specific antitoxins, or by immunological methods such as ELISA or latex agglutination. More recent developments include the application of DNA probe technology, but current methods using DNA probes are neither sufficiently rapid nor cost-effective. The introduction of non-isotopic probes may stimulate further use of DNA probes.

Isolation of obligate anaerobes from clinical material depends very much on the quality of the specimen reaching the laboratory. The most important factor in the survival of anaerobes in transit is the time taken for the specimen to reach the laboratory. The survival of anaerobes on swabs, even in transport

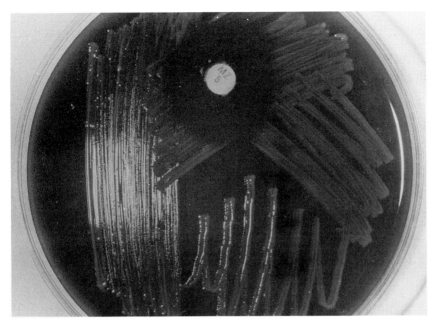

Fig. 6.3 Pure growth of *Bacteroides fragilis* on blood agar, showing sensitivity to metronidazole (5 μg disc).

medium, may be quite brief; specimens of pus or tissue are more likely to yield obligate anaerobes than are swabs. Isolation of obligate anaerobes is facilitated by the use of selective media containing antibiotics (see Chapter 2). After inoculation a 5 μg metronidazole disc is placed on each anaerobic culture plate. After incubation for 48–72 hours, recognition of obligate anaerobes is aided by the resulting zone of inhibition (Fig. 6.3).

Selective media are also required for the isolation of *C. difficile* from faeces of patients with antibiotic-associated diarrhoea. An additional selective procedure for clostridial spores is the use of alcohol shock. A 50% (v/v) suspension of the specimen is made in laboratory alcohol and incubated at room temperature for 1 hour, after which a few drops are plated onto selective and non-selective media. Clostridia may also be isolated from contaminated material following heating in cooked meat broth for 10 minutes at 80°C.

References and further reading

Finegold, S. M. and George, W. L. (eds) (1989) *Anaerobic Infections in Humans.* San Diego: Academic Press.

Lyerly, D. M., Krivan, H. C. and Wilkins, T. D. (1988) *Clostridium difficile*: its disease and toxins. *Clinical Microbiology Reviews* **1**, 1–18.

Smith, L. DS. and Sugiyama, H. (1988) *Botulism: The Organism, its Toxins, the Disease*, 2nd edn. Springfield, Illinois: C. C. Thomas.

Smith, L. DS. and Williams, B. L. (1984) *The Pathogenic Anaerobic Bacteria*, 3rd edn. Springfield, Illinois: C. C. Thomas.

Sugiyama, H. (1980) *Clostridium botulinum* neurotoxin. *Microbiological Reviews* **44**, 419–448.

Sutter, V. L., Citron, D. M., Edelstein, M. A. C. and Finegold, S. M. (1985) *Wadsworth Anaerobic Bacteriology Manual*, 4th edn. Belmont, California: Star Publications.

Willis, A. T. (1969) *Clostridia of Wound Infection*. London: Butterworths.

Willis, A. T. (1977) *Anaerobic Bacteriology: Clinical and Laboratory Practice*, 3rd edn. London: Butterworths.

Willis, A. T. and Phillips, K. D. (1988) *Anaerobic Infections*. London: Public Health Laboratory Service.

7

Industrial importance of obligate anaerobes

Obligate anaerobes are of considerable economic importance in many areas of industrial microbiology. The range of processes in which anaerobes are involved is wide and includes the production of foods, food spoilage and food poisoning, the production of organic solvents and other chemicals by fermentations, anaerobic treatment of sewage, industrial and agricultural wastes, the retting of jute and flax, and corrosion of metals in aquatic environments.

A wide variety of raw food materials is fermented to products which have desirable qualities, one of which is greater stability than the starting material. The majority of these fermentations are performed by lactobacilli. *Lactobacillus plantarum* plays an important role in the pickling of olives, cucumbers and cabbage (sauerkraut), and in the production of silage for use as a winter cattle feed. Sourdough breads also depend for their flavour upon a heterofermentative *Lactobacillus*.

Anaerobes in dairy products

Lactobacilli are employed in the production of a variety of dairy products, often in conjunction with facultative anaerobes such as streptococci and yeasts. These products include cheeses of many types and fermented milk products from around the world (Table 7.1).

Cheeses

Cheese is prepared by adding lactic acid bacteria and rennin, an enzyme preparation obtained from the fourth stomach of unweaned calves, to milk. Lactic acid curdles the milk, while the rennin enzymes hydrolyse milk casein. The polypeptides produced from casein hydrolysis form a lattice after

Table 7.1 Dairy products produced by a lactic fermentation

Food	Species involved
Cheeses	
Emmental, Gruyère	*Lactobacillus bulgaricus, L. lactis, L. helveticus*
Parmesan, Grana	*L. bulgaricus, L. lactis, L. helveticus, L. plantarum, L. acidophilus, L. casei*
Fermented milks	
Bulgarian buttermilk	*L. bulgaricus*
Yakult (Japan)	*L. casei*
Acidophilus milk	*L. acidophilus*
Yoghurt	*L. thermophilus, L. bulgaricus*
Dahi (India)	*L. bulgaricus, L. plantarum*
Kefir (USSR)	*L. acidophilus, L. kefir*
Laban (Lebanon)	*L. bulgaricus*
Koumiss (USSR)	*L. bulgaricus*

reacting with calcium in the milk; this matrix retains the solids from the curdled milk, forming the curd. During the cooking process the curd contracts, expelling the remaining liquid (whey). After cooking, the curd is pressed before storage and ripening. Traditionally the starter cultures for the production of cheese were batches of soured milk or cream with appropriate flavour and acidity. Modern cheese-production plants use starter cultures of appropriate organisms, either in pure or mixed culture.

The cheeses in which lactobacilli are important are all hard cheeses (Table 7.1) produced by a process involving cooking of the curd at high temperatures (50–56°C for 30 minutes). The six *Lactobacillus* spp. employed in the manufacture of these cheeses are all relatively thermophilic and are used in combination with *Streptococcus thermophilus*.

The streptococci develop first, producing L[+]-lactic acid; the lactobacilli grow more slowly, utilizing residual carbohydrate and performing a heterolactic fermentation (D[−] and L[+] lactic acid are the products). Acid production proceeds more rapidly in the outer portions of the cheese as bacterial numbers increase faster due to differential cooling, which establishes a temperature gradient from the centre of the cheese outwards.

The cheeses are salted immediately after production. In the case of Emmental this is accomplished by soaking in brine for 24–48 hours; the Italian Grana cheeses are soaked for up to two weeks. In contrast, dry salt is rubbed into the surface of Gruyère cheeses several times a week. After salting, the cheeses are held in a cool environment (<15°C) for up to a month, while the thermophilic starter flora declines and is replaced by a salt-tolerant flora primarily of propionibacteria (*Propionibacterium freudenreichii* and *P. shermanii*) and yeasts. In Gruyère cheeses a rind develops, while rindless Swiss cheeses are wrapped in a film after soaking in brine, producing a similar effect. Gas exchange between cheese and air is limited and the Eh is greatly reduced, to between -200 and $-300\,$mV. A propionic fermentation by anaerobic propionibacteria develops:

$$3\,[\text{lactate}] \longrightarrow [\text{acetate}] + CO_2 + 2\,[\text{propionate}]$$

This process occurs over a period of up to two months, while the cheeses are stored at a temperature within the range 15–25°C. The numbers of propionibacteria increase to 10^9/g cheese. During this period of storage, growth of propionibacteria is accompanied by the development of the characteristic 'eyes' associated with Swiss cheeses. The 'eyes', containing CO_2, develop first in the centre of the cheese, where growth of propionibacteria is fastest and utilization of lactate is more rapid. This differential growth within the cheese is a function of the low Eh at the centre of the cheese and the higher salt concentration at the periphery.

'Eye' formation is also affected by pH, copper (optimum concentration 14.5 mg/kg cheese) and temperature. Growth of propionibacteria is limited below 15°C; storage of French Beaufort cheeses below 15°C results in a final product with no 'eyes'. The 'eyes' in Gruyère (ripened at 16–18°C) are normally smaller than those in Emmental (ripened at 22°C). When 'eye' formation has reached an optimum stage the cheeses are transferred to a cold room at 10–12°C, where ripening continues over a period of several months, associated with the development of the desired flavour. At this stage some strains of propionibacteria are able to continue metabolizing lactate, resulting in the formation of over-large 'eyes' and occasionally splitting of the cheese. This tendency is minimized by the maintenance of stable conditions throughout the ripening period, with particular attention being paid to temperature control.

Apart from the high-temperature cooked cheeses such as Gruyère and Emmental, 'eyes' are also present to a lesser extent in Dutch Gouda and Edam, New Zealand Egmont, Danish Samsø, Scandinavian Herrgådost and Jarlsbergost, Swiss Appenzell and Italian Fontina cheeses.

Fermented milks

Fermented milks are subdivided into four categories, according to the organisms employed in the fermentation. Bulgarian buttermilk, Yakult and acidophilus milk are all produced by monospecific type II fermentations using lactobacilli (Table 7.1). Acidophilus milk may be a suitable alternative to raw milk for individuals who are lactose-intolerant, since the lactic acid fermentation removes lactose. Yoghurt and Dahi are the result of type-III fermentations employing mixed cultures of lactobacilli and *S. thermophilus*. Type-IV fermentations, such as Kefir, Laban and Koumiss, involve complex mixed cultures of yeasts, lactobacilli and streptococci. The final products of type-IV fermentations all contain a low concentration of alcohol (<3% v/v).

Anaerobic spoilage of dairy products

Most fermented dairy products are not subject to spoilage by anaerobes due to the low pH attained following fermentation. However the potential for spoilage of hard, cooked cheeses due to excess CO_2 production by propionibacteria has already been mentioned. Clostridia also cause spoilage of hard cheeses, known as 'butyric-blowing' or 'late-blowing'.

Milk is contaminated by clostridia at the time of milking. Two species, *C. sporogenes* and *C. tyrobutyricum*, together account for more than 75% of all clostridia in milk. *Clostridium tyrobutyricum* is particularly common in milk from cattle fed silage. The extent of contamination is related to the degree of hygiene practised, but most raw milk contains low numbers of clostridial spores (<200 spores/litre in cattle not fed silage, but 10^3–10^4 spores/litre in cattle fed silage). Clostridia do not usually grow in fresh milk because of the relatively high Eh. Similarly, the growth of clostridia in the majority of dairy products is prevented by high Eh, low pH, low water activity, high salt content and adverse temperature, or a combination of these factors.

Butyric-blowing of hard cheeses is caused by *C. tyrobutyricum*, which is able to multiply at pH 4.5 at the salt and lactic acid concentrations present during ripening. *Clostridium tyrobutyricum* is a saccharolytic organism which ferments lactate, producing acetic and butyric acids, CO_2 and H_2. Gas production results in the formation of holes in the cheese. As the concentration of butyric acid increases the cheese becomes progressively more rancid. Hole formation and rancidity both develop throughout the ripening period.

Butyric-blowing occurs in a variety of cheeses, including Edam, Gouda, Gruyère and Parmesan, but not all cheese varieties are equally susceptible. Cheddar cheese may also be affected, but hole formation does not occur unless the cheese is packaged. It is difficult to predict whether a given batch of cheese will be affected by butyric-blowing. Heavily contaminated milk will be more likely to produce defective cheese; however, even 10^2–10^3 *C. tyrobutyricum* spores/litre milk may be enough to cause the defect in some cheeses, such as Gouda or Emmental. Splitting of the cheese may occur if the raw milk was heavily contaminated with *C. tyrobutyricum* (10^5 spores/litre).

Other clostridia do not generally cause spoilage but may be present as contaminants in cheese.

Methods for prevention of butyric-blowing are related to general hygiene and are aimed at reducing the load of *C. tyrobutyricum* spores in milk. In cheese-producing regions of Switzerland and France dairy cattle are not fed silage, since this is the major source of *C. tyrobutyricum*. Great attention is paid to hygiene during milking. Other factors, including heat treatment of milk and addition of inhibitors such as nitrate/nitrite and lysozyme to cheeses, are of little value because lactobacilli and propionibacteria are more susceptible to these treatments than is *C. tyrobutyricum*. High-speed centrifugation may be helpful in reducing the number of *C. tyrobutyricum* spores in milk prior to cheese manufacture.

Spoilage of non-dairy foods by anaerobes

Meat and fish products

Anaerobes are not generally regarded as a cause of spoilage of meat or fish, despite their presence in tissues of freshly killed animals. One exception is *Clostridium putrefaciens*, a psychrophilic organism which is tolerant of both salt and nitrite. *Clostridium putrefaciens* was recognized as the cause of 'bone-taint' in cured pork products, particularly hams. However, modern methods of mass production have made this problem much less common than was formerly so.

Uneviscerated poultry hung at ambient temperature (15°C) develop greening of the skin within three days, whereas the development of greening takes much longer (up to 18 days) if the birds are hung in a cold store at 5°C. Greening is due to microbial production of H_2S in the intestinal tract. The H_2S diffuses into muscle tissue and under aerobic conditions reacts with haemoglobin to form sulphaemoglobin, giving the characteristic pigmentation. Greening can be delayed by including milk powder in the diet before slaughter of the birds; a lactobacillary flora develops which lowers the caecal pH to <5, inhibiting early growth of sulphide-producing clostridia.

In contrast to the spoilage of poultry caused by intestinal anaerobes, game birds such as grouse, pheasant and partridge are hung, before plucking, for 10–15 days at a temperature of approximately 10°C. During this period, the meat becomes more tender and a characteristic strong flavour develops. This process appears to be the result of metabolism by intestinal anaerobes, of which clostridia are an important component, yet greening does not often occur in game birds.

Vegetables and fruit

Anaerobes may cause spoilage of raw, canned and fermented vegetables and fruit. Because of their ubiquitous distribution in the environment, clostridia are often found on fresh vegetables and fruit. However, numbers of clostridia are usually low and only in potatoes is clostridial spoilage a major problem, although other root vegetables such as carrots may be affected from time to time.

The principal cause of spoilage of potatoes, before and after harvesting, is bacterial soft rot, caused by *Erwinia carotovora*. In this condition the pectic components of the middle lamella and cell wall are degraded by bacterial enzymes. However, several species of clostridia (primarily *C. puniceum*) also produce pectinolytic enzymes and are capable of producing soft rot in potatoes. This occurs particularly in waterlogged soils, when the oxygen concentration surrounding the potato tubers may be very low.

Commercially produced canned vegetables and fruit are heat-treated during processing so that under normal circumstances spoilage and food poisoning organisms are absent from the finished product. However, such treatments are rarely sufficient to eliminate spore-forming bacteria. Growth of organisms that survive processing must be prevented by conditions within the can. The Eh within canned vegetables and fruit is usually negative and may be as low as $-400\,mV$, thus prevention of anaerobic growth depends upon the high acidity of such products. This is particularly important in home-canned and home-bottled vegetables and fruit, which are important vehicles of botulism in the USA (see the section below on food poisoning).

Putrefactive anaerobes such as *C. sporogenes* are often recovered from blown cans of low-acid vegetables, such as baked beans, sweet potatoes, green beans, asparagus and mushrooms. Spores of thermophilic anaerobes such as *C. thermosaccharolyticum* and *Desulfotomaculum nigrificans* may be highly resistant to heat, having D values at 121°C of around three minutes (the D value indicates the time necessary at the stated temperature to bring about a 90% reduction in viable count). These organisms may survive heat treatments and cause spoilage if canned products are exposed to high temperatures. Low-acid vegetables are also more susceptible to spoilage by these two organisms. *Clostridium thermosaccharolyticum* can grow at pH 4.7, whereas *D. nigrificans* will not grow below pH 6. 2. Spoilage by these organisms is a greater problem in the tropics than in temperate regions; the optimum growth temperatures for both *C. thermosaccharolyticum* and *C. nigrificans* are 55°C, but they will grow at temperatures as low as 37°C and 43°C, respectively.

Mesophilic, non-proteolytic, butyrate-producing organisms such as *C. butyricum* and *C. pasteurianum* are less heat-resistant than *C. sporogenes*, but they are a significant cause of spoilage in more acid foods (pH <4.5) such as fruits, which receive a less stringent heat treatment than low-acid foods.

Prevention of spoilage of canned fruit and vegetables by spore-forming anaerobes depends on rigorous maintenance of the integrity of the process and upon reducing the numbers of spores on raw material to an absolute minimum by washing (optimum 1 spore/10 g raw material).

Numerous vegetables and fruits are preserved by fermentation in brine. The salt concentration used depends on the particular fruit or vegetable to be preserved. Under appropriate conditions a lactic acid fermentation occurs and the final pH of the product is usually <4, which is sufficient to inhibit most spoilage organisms. Air is excluded to prevent the growth of moulds and yeasts. Common examples of vegetables preserved in this way are cabbage, olives, cucumbers, peppers and green tomatoes.

Several butyrate-producing clostridia are important causes of spoilage, particularly of green olives. Two types of spoilage occur, the first taking place before the development of the lactic acid fermentation. This form of spoilage is associated with production of butyric acid in the brine and bubbles of hydrogen in the olive tissues. Spoilage of this type is invariably caused by *C. beijerinckii/butyricum* strains capable of growth at pH 4.5 and in the presence of 5% NaCl. In contrast, 'zapatera' spoilage occurs later in the preservation process, if the pH of the brine remains above 4.5 after the cessation of the lactic fermentation. This type of spoilage is accompanied by the development of a putrefactive odour, due to the growth of *C. bifermentans* or *C. sporogenes*.

Food poisoning

Clostridial food poisoning caused by enterotoxigenic *C. perfringens* type A is outside the scope of this discussion and is discussed in Chapter 6. However the occurrence and potential growth of *C. botulinum* in foods is of great significance to the food industry. *Clostridium botulinum* is widely distributed in soil and in some foods. Much effort is directed towards the prevention of germination of *C. botulinum* spores that remain in processed foods.

Traditional heat processing involves the packing of non-sterile foods into non-sterile containers which are then sealed and treated with steam at temperatures from 110–126°C. Ultra-high temperature (UHT) processing entails the application of temperatures up to 150°C. For products in which *C. botulinum* can grow, the minimum heat process is designed to reduce the probability of survival of spores to less than 1 in 10^{12}. Such a process is usually defined as being equivalent to three minutes at 121°C (an F_o3 process).

Foods which are not suitable for such rigorous heat treatment must be preserved in other ways. Heat-labile meat products may be preserved by the addition of salts such as sodium chloride and sodium nitrite. Both compounds inhibit specific enzymes in *C. botulinum*. The combination of these two compounds is synergistic.

Foods which are not naturally acidic can be acidified to <pH 4.6, thus providing a high degree of safety against the growth of *C. botulinum*. Increasingly, combined physical and chemical factors are employed in order to render food products safe. One example is the use of sub-optimum pH, suboptimum temperature (during storage) and acidification with citric acid. At pH 5.2, citric acid prevents growth of *C. botulinum* by chelating divalent cations, particularly calcium. As the food-manufacturing industry continues to produce new food products, the development and application of inhibitors of *C. botulinum* will increasingly be of significance.

Anaerobic digestion

Anaerobic digestion of organic matter to methane is a widespread process in natural environments (see Chapter 4). Methane production is a syntrophic process depending upon the action of several types of anaerobic bacteria. There are four steps in this process (Fig. 7.1). The initial stage of hydrolysis is performed by a variety of organisms, chiefly clostridia, which may reach counts of up to 10^9 cells/ml in the digester contents. The major intermediary products of the second stage, acidogenesis, are short-chain fatty acids, hydrogen and carbon dioxide. The third stage (acetification) is the result of metabolism of fatty acids by H_2-producing acetogenic bacteria. These organisms are unable to grow at partial pressures of hydrogen >10^{-3}atm., thus their maintenance within the methanogenic consortium depends upon the continued removal of hydrogen by methanogens. This is known as interspecies hydrogen transfer (see Chapter 5). Because of their syntrophic relationship with the methanogens, H_2-producing acetogens are difficult to isolate in pure culture. However, some species have been isolated and characterized, including *Syntrophobacter wolinii* and *Syntrophomonas wolfei*. Acetogens of this type reach densities of 10^6–10^7 cells/ml in sludge. Some hydrogen is converted to acetate by H_2-consuming acetogens (homoacetogens), such as *Acetobacterium woodii*, *Acetogenium kivui*, *Clostridium thermautotrophicum* and *C. formicaceticum*.

About 70% of the methane generated by the anaerobic digestion of organic matter is produced from acetate by the acetoclastic methanogens; the remainder is derived from H_2 and CO_2 by the action of hydrogenotrophic methanogens. The acetoclastic methanogens are very slow growing and thus a high retention time is necessary for maximum methane production. The gas evolved during anaerobic digestion is a mixture of approximately 65%

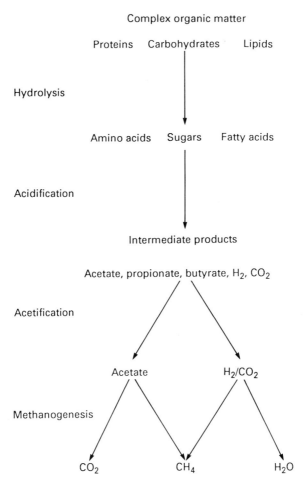

Fig. 7.1 Stages in the anaerobic digestion of organic matter.

methane and 35% carbon dioxide and is known by a variety of synonyms, such as 'marsh gas', 'sewer gas' and more recently 'biogas'.

Within digesters, methanogenesis occurs at an optimum rate in the range pH 6–8. The optimum temperature for mesophilic digestion is 40°C, whereas thermophilic digesters have an optimum temperature of 60°C. Methanogens are susceptible to inhibition by a variety of organic pollutants (including CCl_4, $CHCl_3$, CH_2Cl_2 and CN) and heavy-metal ions, at concentrations as low as 1 ppm. Formaldehyde, SO_2 and H_2S are also toxic at higher concentrations (50–400 ppm). In addition, ammonia may be inhibitory unless time is allowed for the digester flora to adapt to higher ammonia levels.

Table 7.2 Applications of anaerobic digestion

● Treatment of:	human sewage
	farm slurry
	domestic refuse
● Disposal of:	crop residues
	bagasse
	palm-oil waste
	straw
	food-processing wastes
	dairies
	breweries
	slaughterhouses
	olive-oil mills
	wool textile industry waste
	papermill sludge
● Production of biogas from agricultural crops	

This is of particular relevance when the feedstock is porcine excrement or chicken litter, both of which are high in ammonia.

Applications of anaerobic digestion

Research on anaerobic digestion of human sewage began in the latter half of the 19th century. The septic tank was developed in England by Donald Cameron in 1895. The septic tank is basically a sealed chamber which is allowed to fill with effluent, in which an anaerobic fermentation develops. Septic tanks remain at ambient temperature, so digestion of sewage is a slow process in temperate climates and relatively little methane is evolved. The resulting liquid effluent may be discharged safely onto land, while the remaining sludge has to be cleared from the tank at intervals. The first municipal plant for anaerobic treatment of sewage was established in Exeter in 1897, using the septic-tank principle. Methane from the plant was used to provide heating and lighting within the sewage works.

Anaerobic digestion is widely used as a treatment for organic wastes (Table 7.2). The feedstock may be excrement of human or animal origin, domestic refuse in landfill sites, residues from agriculture and food-processing industries or agricultural crops grown specifically for the purpose of biogas generation. Anaerobic digestion of wastes offers a number of benefits (Table 7.3).

Removal of pollutants by anaerobic digestion is comparable with the best aerobic treatment systems. BOD may be reduced by approximately 80%, COD by 50% and the total solids by 40%. Despite the great reduction in

Table 7.3 Benefits of anaerobic digestion

● Pollution control:	reduction of:
	BOD
	COD
	bacterial pathogens
	weed seeds
	odour
● Production of:	fuel (biogas)
	fertilizer
	composted solids
	liquids

polluting power, the effluent from anaerobic digestion is still unfit for discharge into waterways and thus usually requires some secondary treatment such as settling or aeration. An alternative is the separation of liquid and solid fractions. The solid residue of digestion contains about 25% protein–nitrogen, 3% ammonia–nitrogen, 9% fat, 6% phosphate and 1.4% potassium. After stabilization by composting, it is thus a highly nutrient-rich fertilizer. The liquid fraction of the effluent may be sprayed directly onto fields used for grazing, since it has little odour and numbers both of bacterial pathogens and of weed seeds are considerably reduced. Thus the period for which the fields cannot be grazed is much shorter than is so when conventional slurry spreading is practised.

One area of special interest is the ability of methanogenic consortia to degrade aromatic compounds, such as benzene and toluene:

$$\text{toluene} + H_2O \longrightarrow p\text{-cresol} + H_2 \ (\Delta G^{\circ\prime} = +71 \text{ kJ/mole})$$
$$\text{benzene} + H_2O \longrightarrow \text{phenol} + H_2 \ (\Delta G^{\circ\prime} = +73 \text{ kJ/mole})$$

These are endergonic reactions, which are coupled to reductive dehydroxylation or dehalogenation as energy-yielding reactions. Methanogenic consortia metabolize benzoate by reduction, followed by β-oxidation, leading to ring cleavage and the production of fatty acids which serve as substrates for acetification and methanogenesis. The overall reaction is:

$$4C_6H_5COOH + 18H_2O \longrightarrow 15CH_4 + 13CO_2$$

Aromatic compounds are initially metabolized only slowly within a digester. However, adaptation of the flora to aromatic substrates occurs after 3–4 weeks. After adaptation has occurred, complete digestion takes only a few days. One group of aromatic compounds which are almost completely resistant to anaerobic degradation by methanogenic consortia are the lignified tissues of plants.

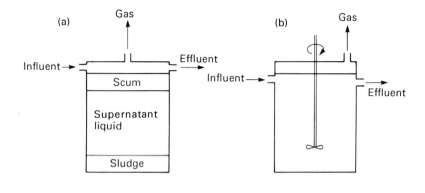

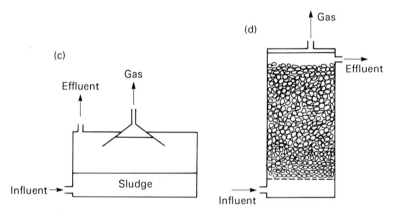

Fig. 7.2 Anaerobic digester designs: (a) conventional unstirred digester; (b) high-rate stirred digester; (c) upflow anaerobic sludge blanket (UASB) digester; (d) anaerobic stationary-bed fixed film digester.

Anaerobic digesters are operated throughout the world. It is estimated that in China there are several million small domestic and community digesters in operation. Digestion is widely utilized in India both for treatment of sewage and for generation of fuel methane. In Europe and the USA, the emphasis has been on the treatment of wastes rather than of fuel generation. The methane produced has been regarded as a by-product, since it has not been commercially viable to use anaerobic digestion as a means of generating fuel. However, in many agricultural applications the use or sale of liquid and solid effluent fractions as fertilizers has made the installation of anaerobic digesters financially attractive. Most farm-based digesters in Europe are $<1000\,m^3$, serving populations up to an equivalent of 25 000 pigs or 50 000 humans. A number of municipal sewage treatment digesters in the UK are up to 5000 m³. Digesters of this size may produce as much as 55 000 m³ gas/day.

Types of anaerobic digester

Early digesters resembled septic tanks in design. Most traditional digesters were constructed in the form of upright cylindrical tanks (Fig. 7.2), into which the feedstock was pumped and from which the effluent was removed, also by pumping. Gas was vented from the top of the cylinder. In this type of digester very little mixing occurred, and a distinct sludge layer settled at the bottom of the digester. A thick layer of scum formed on top of the liquid fraction, which often restricted the working volume of the digester.

The introduction of stirred digesters (Fig. 7.2) resolved the problems of sludge settling and scum formation and in addition produced a relative increase in methanogenic activity. The increased metabolic turnover is due to better mixing of the methanogenic flocs with substrate. Most digesters used for treatment of human or animal wastes are now of this basic design.

The use of anerobic digestion as a treatment for low-strength, high-volume wastes such as those from food-processing plants (Table 7.2) causes the problem of washout of the methanogenic consortia from the digester, because of the high hydraulic loading rates necessary when the substrate is very dilute. One remedy for this problem is the immobilization of the methanogenic flora on solid particles, in the fixed-film digester (Fig. 7.2). In this type of digester, the influent is pumped up through a column of granules on which the methanogenic consortia develop. It was soon found that the settling properties of the methanogenic flocs were such that the solid particles in fixed film digesters were in many cases superfluous. A recent development is the upflow anaerobic sludge blanket (UASB) digester (Fig. 7.2). In this design the sludge layer is allowed to settle and the influent is pumped up through the sludge from the bottom of the digester. Mixing occurs as the gas is produced and rises through the sludge.

In all the digesters described above, the entire process of digestion takes place within one chamber (Fig. 7.3). There are certain disadvantages associated with conventional (one-stage) digester design. One-stage digesters are susceptible to the effects of substrate overloading. When this occurs the digester pH falls as volatile fatty acids accumulate in excess, and both acetogenesis and methanogenesis are inhibited. This imbalance can be corrected by stopping the flow of substrate into the digester until the flora has equilibrated and methanogenesis recommences. Inhibition due to overloading is costly both in terms of labour and lost methane production.

One approach which has been adopted to overcome this and other problems is the construction of two-stage digesters (Fig. 7.3). In digesters of

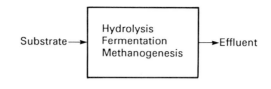

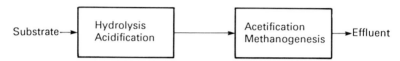

Fig. 7.3 Principle of single-stage (upper) and two-stage (lower) anaerobic digester design.

this design, hydrolysis and acidification are separated from acetification and methanogenesis. The two-stage digester has several potential advantages:

(i) it allows optimum operation of both phases of digestion;
(ii) it prevents inhibition of methanogenesis due to substrate overloading;
(iii) it is less susceptible to changes in feedstock composition.

However, the two-stage process is more complex (and therefore more costly) both to establish and to control.

Anaerobic retting of plant fibres

Several stem fibre crops are macerated by microbial enzymes during the process known as 'retting'. These include jute (*Corchorus* spp.), flax (*Linum usitatissimum*) and hemp (*Cannabis sativa*). During retting the bast fibres are released from the stem following maceration of the surrounding cortex. After retting, straw is dried and the fibres are freed from stem debris by a mechanical process known as 'scutching'.

There are two processes which are used for commercial retting. Dew retting occurs when the plants are spread thinly on the ground after pulling and allowed to ret without further manipulation. During this process, which takes between two weeks and two months, a flora develops which is composed primarily of pectinolytic fungi and aerobic bacteria. The successful application of dew retting depends upon both a reasonably warm climate and low rainfall, since repeated soaking leads to rotting of the straw.

The second process is known as 'tank- or dam-retting'. Pulled plants are bundled and stacked together, before being submerged in water for up to five days at a temperature of 30–37°C. In temperate regions the water has to be heated whereas in tropical countries the ret is conducted at ambient

Table 7.4 Production of chemicals by anaerobic fermentations (*Clostridium*)

Product	Species responsible
Acetic acid	*C. aceticum, C. formicaceticum, C. thermaceticum, C. thermautotrophicum*
Acrylic acid	*C. propionicum*
Butyric acid	*C. butyricum, C. cellulovorans, C. pasteurianum, C. thermosaccharolyticum*
Propionic acid	*C. propionicum*
Acetone/butanol	*C. acetobutylicum, C. aurantibutyricum, C. beijerinckii, 'C. tetanomorphum'*
Ethanol	*C. saccharolyticum, C. thermocellum, C. thermohydrosulfuricum*

temperatures. The source of the retting water is considered important and is often a small spring. In traditional dam-retting, pits are flooded with river water. During tank-retting, an anaerobic flora develops, of which clostridia are important components. *Clostridium felsineum* is thought to be the most important anaerobe in tank retting, but in addition *C. butyricum* and *C. acetobutylicum* reach high numbers on the surface of, and within, retting straw. *Bacillus subtilis* also exhibits retting activity in laboratory scale, pure culture rets.

Pectins within the straw are degraded by three types of enzyme: polygalacturonases, pectin lyases and pectin esterases. Of the organisms which have good retting activity, *C. felsineum* has been most extensively studied. *Clostridium felsineum* produces polygalacturonase and pectin lyase, whereas these enzymes have not been detected from other retting organisms.

Production of solvents and other chemicals

A number of organic solvents and other chemicals can be produced by anaerobic fermentations. Before the development of the petrochemical industry, some of these fermentations were economically viable and were conducted on a large scale throughout the industrialized world. The organisms responsible for these fermentations are exclusively clostridia (Table 7.4).

Much of the early interest in the acetone/butanol fermentation centred on the potential of butanol as a source of butadiene, a precursor of synthetic rubber. During the First World War the emphasis turned to the production of acetone, for use in the manufacture of cordite. Butanol was regarded as a waste product. After the war, butanol was again in demand as a solvent in nitrocellulose lacquers for the motor industry.

Initially maize was used as the substrate in the Weizmann process. In the 1930s molasses became the most widely used substrate. After the Second World War microbial production of butanol became relatively more costly than production by the petrochemical route. The rising costs of oil since the 1970s have prompted renewed interest in the acetone/butanol fermentation.

The industrial-scale process currently in use is a batch fermentation. The inoculum is prepared by serial subculture of a heat-shocked spore suspension in progressively larger volumes of medium, until approximately 30 litres is inoculated into a fermenter containing 90 000 litres of invert molasses supplemented with an additional source of nitrogen, usually corn steep liquor or yeast water. The optimum pH range is 5.5–6.5 and the optimum temperature for the fermentation is between 29 and 35°C. The incubation period is usually 50–55 hours. The ratio of solvents produced by this process is 75% butanol:20% acetone:5% ethanol (compared with 60% butanol:30% acetone:10% ethanol from the Weizmann process using a maize substrate). Cooling of the fermentation after 16 hours to 24–25°C increases the proportion of butanol produced.

The major limitation to the microbial production of acetone and butanol is the cost of recovering the products from the spent medium by distillation, since the total volume of solvents is 2–3%. Recovery by distillation accounts for two-thirds of the direct energy costs of the entire process. Much of the remaining cost is associated with the sterilization of the fermentation vessel before each batch is processed.

Recent research has concentrated on maximizing the yield of solvents. Alternative substrates such as cheese whey, waste sulphite liquor, bagasse and rice straw have been investigated. In addition, much effort has been directed towards strain improvement as a means of increasing yields. One of the limiting factors in the butanol fermentation is the induction of autolysis by butanol at concentrations of 7–16 g/litre. However, autolysis-deficient mutants have been isolated which can continue growth in higher concentrations of butanol, thus increasing the yield/substrate ratio. Other approaches have included the selection of butanol-resistant mutants, which also produce more butanol than the parent strains.

The application of continuous culture techniques to the acetone/butanol fermentation has yet to be successful on the industrial scale. Among the problems encountered have been contamination, mutation and selection of strains with undesirable characteristics. The loss of solvent-producing ability is particularly common in continuous culture systems. Use of immobilized cells or spores of *C. acetobutylicum* currently appears to confer little advantage over conventional batch fermentation. However, laboratory-scale

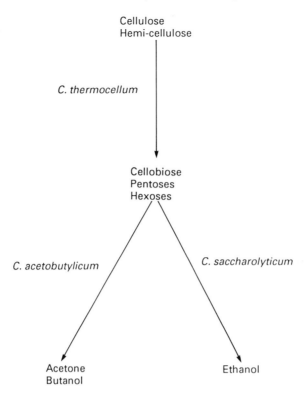

Fig. 7.4 Use of co-cultures for production of solvents from cellulose. Products may be varied by use of alternative non-cellulolytic organisms.

continuous culture processes, using cells of *C. acetobutylicum* immobilized on bonechar, and whey permeate as the substrate, have been developed. This area is one of particularly active research and scale-up to industrial size fermentations may be possible in the future.

The remaining clostridial fermentations (Table 7.4) are presently of academic rather than industrial interest. However, research on the production of ethanol and several organic acids continues and in the future the industrial use of such fermentations may become economically viable.

Two of the ethanol-producing clostridia, *C. thermocellum* and *C. thermohydrosulfuricum*, are thermophiles with temperature optima between 65 and 72°C. While this characteristic confers advantages in terms of increased production of ethanol, the costs of operating fermenters at the optimum temperature are also higher. *Clostridium thermocellum* produces cellulolytic enzymes but metabolizes a limited range of other substrates, with acetic acid and ethanol as end-products. In contrast, *C. thermohydrosulfuricum* and

C. saccharolyticum are metabolically versatile but are not cellulolytic; they produce ethanol and CO_2 as their major end-products.

Much interest has been directed towards the use of co-cultures of a cellulolytic organism (such as *C. thermocellum* or *C. cellulovorans*) and another species with a desired end-product (Rogers, 1986). Such a combination allows the use of cellulose-containing materials as substrates for the production of a wide range of compounds and is of great industrial potential. An example of this approach is illustrated in Fig. 7.4. The limiting factor is the requirement for the non-cellulolytic organism to be capable of fermenting cellobiose, which otherwise accumulates and inhibits the cellulolytic enzyme complexes.

The homoacetogenic clostridia are potentially useful because they produce 3 moles acetate from 1 mole glucose. This is because of their ability to fix CO_2:

$$2CO_2 + 4H_2 \longrightarrow CH_3COOH + 2H_2O$$

Some acetogenic species can also synthesize acetate from one-carbon compounds such as methanol and formate; *C. thermaceticum* and *C. thermautotrophicum* can utilize carbon monoxide:

$$4CO + 2H_2O \longrightarrow CH_3COOH + 2CO_2$$

Propionic acid is the major end-product of fermentation of α-alanine, β-alanine or lactate by *C. propionicum*. In the presence of oxygen, resting cells of *C. propionicum* accumulate acrylic acid, normally an intermediate in the production of propionic acid. The butyrate-producing clostridia are also of interest. *Clostridium thermosaccharolyticum* switches from production of volatile acids to production of ethanol if cell division is interrupted; *C. cellulovorans* is also of potential value for the conversion of cellulose to butyrate.

Corrosion of metals by anaerobes

The corrosion of metals by microbial action accounts for between one-half and three-quarters of all failures of underground pipelines. Corrosion of underground structures by microorganisms is almost invariably the result of the formation of microbial communities of which sulphate-reducing bacteria (SRB) are the principal components. The costs of such damage are difficult to estimate but certainly run into billions of dollars annually. Other areas in which corrosion by SRB is of great industrial importance include the oil industry (in particular, offshore drilling platforms), chemical manufacturing plants and the shipping industry.

The ecology of surface communities involving SRB was discussed in the

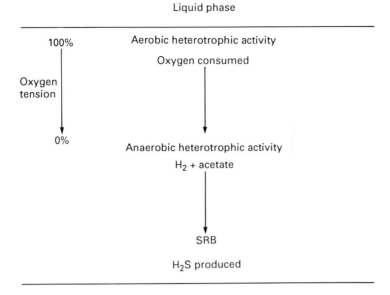

Fig. 7.5 Stratification of surface biofilms in fluid environments (Hamilton, 1985).

context of corrosion by Hamilton (1985). The stratification of surface biofilms in aqueous or other liquid environments is illustrated in Fig. 7.5. SRB form the innermost layer of bacteria in a heterogeneous community.

Corrosion induced by SRB occurs in the absence of oxygen and at or near pH 7. Among the corrosion products formed are iron sulphides. A number of hypotheses have been made to explain the role of SRB in metal corrosion. The majority of these hypotheses are based upon the theory of cathodic depolarization. In this process hydrogen is removed by the hydrogenase of SRB:

dissociation of water: $8H_2O \longrightarrow 8OH^- + 8H^+$

anode: $4Fe \longrightarrow 4Fe^{2+} + 8e^-$

cathode: $8H^+ + 8e^- \longrightarrow 8H$

cathodic
depolarization: $SO_4^{2-} + 8H \longrightarrow S^{2-} + 4H_2O$

anode: $Fe^{2+} + S^{2-} \longrightarrow FeS$ ⎱ corrosion

anode: $3Fe^{2+} + 6OH^- \longrightarrow 3Fe(OH)_2$ ⎰ products

overall reaction: $4Fe + SO_4^{2-} + 4H_2O \longrightarrow FeS + 3Fe(OH)_2 + 2OH^-$

However, some evidence suggests that hydrogenase-dependent cathodic depolarization may not fully explain the action of SRB in corrosion. It has been proposed that the cathodic reactant is H_2S:

$$H_2S + e^- \rightleftharpoons HS^- + \tfrac{1}{2} H_2$$

This hypothesis maintains the importance of hydrogenase, which removes molecular hydrogen and thus generates H_2S, and also recognizes the role of ferrous sulphide as a corroding factor.

Sulphur may also be involved in corrosion by SRB. Indeed corrosion involving SRB proceeds more rapidly in the presence of low concentrations of oxygen. Whether the oxidation of sulphide to sulphur occurs by chemical action, or is performed by aerobic sulphur-oxidizing bacteria such as *Thiobacillus*, is uncertain.

Other problems caused by growth of SRB include blockage of pipes by colloidal precipitates of FeS, souring of oil and gas by H_2S (giving rise to increased SO_2 production on combustion) and poisoning of workers by pockets of H_2S. Attempts to prevent corrosion by SRB have been made in closed systems in the oil and chemical industries using biocides. Particular success has been achieved using quaternary ammonium compounds and isothiazolones. Aeration of aquatic systems is also effective, but may increase the rate of sulphur production and stimulate growth of *Thiobacillus* spp., both of which are undesirable.

Future prospects

The future use of anaerobes as industrial organisms will lie in two major areas – waste treatment, and the manufacture of fuels and solvents. The microbiology of these processes is already well understood and so major improvements in productivity will most probably arise from engineering devlopments, exemplified by the continuing improvements in anaerobic digester design. The contributions of microbiology to these industrial processes are to make the bacteria employed as efficient as possible at the required task(s) and also to ensure that the organisms employed are sufficiently robust to withstand handling on an industrial scale by non-microbiologists.

These challenges can both be met by genetic improvement of industrial strains. In most applications, the strains employed are already the most efficient that can be selected using conventional techniques. One area where this is not the case is the retting of flax in the production of linen. Research is continuing to define the optimum retting flora.

Molecular geneticists have turned their attention to obligate anaerobes only recently. Most research to date has been done on methanogens and the industrial clostridia, principally *C. acetobutylicum*. Despite the fundamental differences between Archaebacteria and Eubacteria, the genes of methanogens closely resemble those of Eubacteria and functional expression of cloned methanogen genes has been reported.

In clostridia, the development of recombinant DNA systems has centred on the use of protoplast transformation. In addition, co-integration using plasmids has been used to transfer genes between saccharolytic clostridia such as *C. butyricum* and *C. acetobutylicum*.

Strain improvement may be of particular value if the range of substrates available to industrial strains can be increased, by inserting genes from other organisms. Some examples include the insertion of the ability to utilize aromatic and/or toxic compounds into methanogens, for use in anaerobic digesters, and the transfer of cellulolytic genes to clostridia with a desired end-product profile. The prospects for improvement of the rumen fermentation in this way were discussed recently (Patterson, 1989).

References and further reading

Archer, D. B. and Thompson, L. A. (1987) Energy production through the treatment of wastes by micro-organisms. *Journal of Applied Bacteriology* **63**, 59S–70S.

Bergère, J.-L. and Accolas, J. -P. (1986) Non-sporing and sporing anaerobes in dairy products. In: *Anaerobic Bacteria in Habitats Other than Man* (eds Barnes, E. M. and Mead, G. C.), pp. 373–396. Oxford: Blackwell Scientific Publications.

Chesson, A. (1980) Maceration in relation to the post-harvest handling and processing of plant material. *Journal of Applied Bacteriology* **48**, 1–45.

Evans, M. R. and Smith, M. P. W. (1986) Treatment of farm animal wastes. *Journal of Applied Bacteriology* **61**, 27S–41S.

Hamilton, W. A. (1985) Sulphate-reducing bacteria and anaerobic corrosion. *Annual Review of Microbiology* **39**, 195–217.

Haylock, R. W. and Donaghy, J. A. (1990) Anaerobic macerations for the production of textile fibres. *Biodeterioration Abstracts* **4**, 97–107.

Iverson, W. P. (1987) Microbial corrosion of metals. *Advances in Applied Microbiology* **32**, 1–36.

Large, P. J. (1983) *Methylotrophy and Methanogenesis*. Wokingham: Van Nostrand Reinhold.

Lund, B. M. (1986) Anaerobes in relation to foods of plant origin. In *Anaerobic Bacteria in Habitats Other than Man* (eds. Barnes, E. M. and Mead, G. C.), pp. 351–372. Oxford: Blackwell Scientific Publications.

McNeil, B. and Kristiansen, B. (1986) The acetone butanol fermentation. *Advances in Applied Microbiology* **31**, 61–92.

Odelson, D. A., Rasmussen, J. L., Smith, C. J. and Macrina, F. L. (1987) Extrachromosomal systems and gene transmission in anaerobic bacteria. *Plasmid* **17**, 87–109.

Patterson, J. A. (1989) Prospects for establishment of genetically engineered microorganisms in the rumen. *Enzyme and Microbial Technology* **11**, 187–189.

Prentice, G. A. and Neaves, P. (1986) The role of micro-organisms in the dairy industry. *Journal of Applied Bacteriology* **61**, 43S–57S.

Rogers, P. (1986) Genetics and biochemistry of *Clostridium* relevant to development of fermentation processes. *Advances in Applied Microbiology* **31**, 1–60.

Sleat, R. and Robinson, J. P. (1984) The bacteriology of anaerobic decomposition of aromatic compounds. *Journal of Applied Bacteriology* **57**, 381–394.

Wachenheim, D. E. and Patterson, J. A. (1988) Potential for industrial polysaccharides from anaerobes. *Enzyme and Microbial Technology* **10**, 56–57.

Zeikus, J. G. (1980) Chemical and fuel production by anaerobic bacteria. *Annual Review of Microbiology* **34**, 423–464.

Index